Sensors and Actuators: An Engineering Perspective

Sensors and Actuators: An Engineering Perspective

Thomas Murphy

Willford Press
www.willfordpress.com

Published by Willford Press,
118-35 Queens Blvd., Suite 400,
Forest Hills, NY 11375, USA

ISBN: 978-1-64728-328-5

Cataloging-in-Publication Data

Sensors and actuators : an engineering perspective / Thomas Murphy.
p. cm.
Includes bibliographical references and index.
ISBN 978-1-64728-328-5
1. Detectors. 2. Actuators. 3. Automotive sensors. 4. Automatic control.
5. Automobiles--Electric equipment. I. Murphy, Thomas.
TK7872.D48 S46 2022
681.2--dc23

For information on all Willford Press publications
visit our website at www.willfordpress.com

Willford Press

TABLE OF CONTENTS

PREFACE

The purpose of this book is to help students understand the fundamental concepts of this discipline. It is designed to motivate students to learn and prosper. I am grateful for the support of my colleagues. I would also like to acknowledge the encouragement of my family.

Sensors are the devices or subsystems that are used for the detection of events and changes in its environment. They collect the information and send it to other electronics. They are used in the objects of everyday use such as touch-sensitive elevator buttons. Two of the common types of sensors include chemical sensors and biosensors. Chemical sensors are the analytical devices that are able to provide information about the chemical composition of their environment. Biosensors are the sensors that detect analytes through biological components such as cell, proteins and nucleic acids. Actuators are the constituents of machines that are responsible to move and control a mechanism. This book presents the complex aspects of sensors and actuators in the most comprehensible and easy to understand language. It aims to shed light on some of the unexplored aspects of such fields. This textbook will provide comprehensive knowledge to the readers.

A foreword for all the chapters is provided below:

Chapter – Introduction to Sensors and Actuators

A sensor refers to device, machine, module or subsystem. Its purpose is to detect events and changes in its environment and to forward the information to other electronics. An actuator is a component of machine which is necessary to move and control a mechanism or system. This chapter will introduce briefly all the significant aspects of sensors and actuators.

Chapter – Transducers

Transducers are the devices that convert energy from one form to another. There are various types of transducers such as resistive transducers, capacitive transducers, piezoelectric transducers, pressure transducers, capacitive micromachined ultrasonic transducers, electromagnetic acoustic transducers and orthomode transducers. All these types of transducers have been carefully analyzed in this chapter.

Chapter – Types of Sensors

Some of the different types of sensors are humidity sensors, biosensors, optical sensors, position sensors, light sensors, Hall effect sensors, oxygen sensors, wheel speed sensors, torque sensors, throttle position sensors, soil moisture sensors, active pixel sensors, etc. This chapter discusses in detail all these types of sensors.

Chapter – Applications of Sensors

Sensors are applied in a number of fields including medical science, transport and logistics, remote sensing, collocation, etc. This chapter has been carefully written to provide an easy understanding of these applications of sensors as well as the use of optical chemical sensors in industries.

Chapter – Types and Applications of Actuators

There are numerous types of actuators that are applied in different areas. Some of them include linear solenoid actuator, pneumatic and hydraulic actuators, carbon nanotube actuators, piezoelectric actuator, helical band actuator, MEMS magnetic actuator and mini/micro shape memory actuators. The topics elaborated in this chapter will help in gaining a better perspective about these types and applications of actuators.

Thomas Murphy

Introduction to Sensors and Actuators 1

- Sensor
- Actuator

A sensor refers to device, machine, module or subsystem. Its purpose is to detect events and changes in its environment and to forward the information to other electronics. An actuator is a component of machine which is necessary to move and control a mechanism or system. This chapter will introduce briefly all the significant aspects of sensors and actuators.

Sensor

A sensor is a device that detects and responds to some type of input from the physical environment. The specific input could be light, heat, motion, moisture, pressure, or any one of a great number of other environmental phenomena. The output is generally a signal that is converted to human-readable display at the sensor location or transmitted electronically over a network for reading or further processing. all types of sensors can be classed as two kinds, either Passive Sensors or Active Sensors.

Generally, active sensors require an external power supply to operate, called an excitation signal which is used by the sensor to produce the output signal. Active sensors are self-generating devices because their own properties change in response to an external effect producing for example, an output voltage of 1 to 10v DC or an output current such as 4 to 20mA DC. Active sensors can also produce signal amplification.

A good example of an active sensor is an LVDT sensor or a strain gauge. Strain gauges are pressure-sensitive resistive bridge networks that are external biased (excitation signal) in such a way as to produce an output voltage in proportion to the amount of force and/or strain being applied to the sensor.

Unlike an active sensor, a passive sensor does not need any additional power source or excitation voltage. Instead a passive sensor generates an output signal in response to some external stimulus. For example, a thermocouple which generates its own voltage output when exposed to heat. Then passive sensors are direct sensors which change their physical properties, such as resistance, capacitance or inductance etc.

But as well as analogue sensors, Digital Sensors produce a discrete output representing a binary number or digit such as a logic level “0” or a logic level “1”.

Analogue and Digital Sensors

Analogue Sensors

Analogue Sensors produce a continuous output signal or voltage which is generally proportional to the quantity being measured. Physical quantities such as Temperature, Speed, Pressure, Displacement, Strain etc are all analogue quantities as they tend to be continuous in nature. For example, the temperature of a liquid can be measured using a thermometer or thermocouple which continuously responds to temperature changes as the liquid is heated up or cooled down.

Thermocouple used to Produce an Analogue Signal

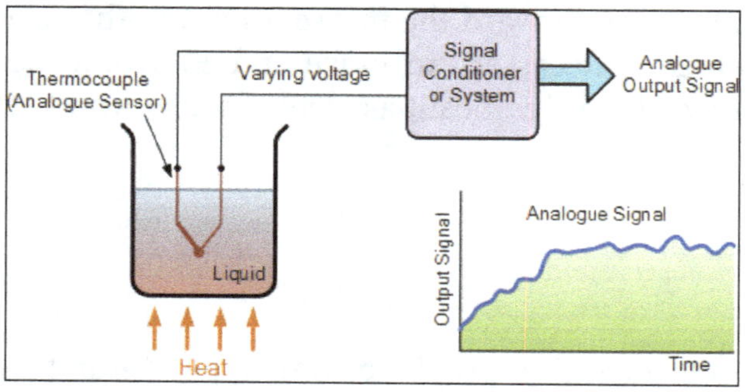

Analogue sensors tend to produce output signals that are changing smoothly and continuously over time. These signals tend to be very small in value from a few mico-volts (uV) to several milli-volts (mV), so some form of amplification is required.

Then circuits which measure analogue signals usually have a slow response and/or low accuracy. Also analogue signals can be easily converted into digital type signals for use in micro-controller systems by the use of analogue-to-digital converters, or ADC’s.

Digital Sensors

As its name implies, Digital Sensors produce a discrete digital output signals or voltages that are a digital representation of the quantity being measured. Digital sensors produce a Binary output signal in the form of a logic “1” or a logic “0”, (“ON” or “OFF”). This means then that a digital signal only produces discrete (non-continuous) values which may be outputted as a single “bit”, (serial transmission) or by combining the bits to produce a single “byte” output (parallel transmission).

Light Sensor used to Produce a Digital Signal

The speed of the rotating shaft is measured by using a digital LED/Opto-detector sensor. The disc which is fixed to a rotating shaft (for example, from a motor or robot wheels), has a number of transparent slots within its design. As the disc rotates with the speed of the shaft, each slot passes by the sensor in turn producing an output pulse representing a logic “1” or logic “0” level.

1

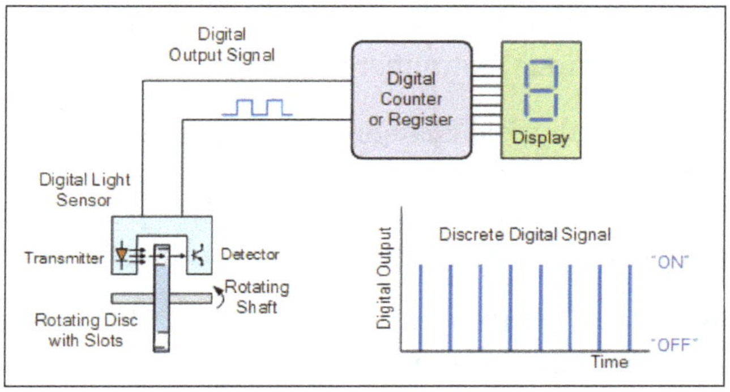

These pulses are sent to a register of counter and finally to an output display to show the speed or revolutions of the shaft. By increasing the number of slots or "windows" within the disc more output pulses can be produced for each revolution of the shaft. The advantage of this is that a greater resolution and accuracy is achieved as fractions of a revolution can be detected. Then this type of sensor arrangement could also be used for positional control with one of the discs slots representing a reference position.

Compared to analogue signals, digital signals or quantities have very high accuracies and can be both measured and "sampled" at a very high clock speed. The accuracy of the digital signal is proportional to the number of bits used to represent the measured quantity. For example, using a processor of 8 bits, will produce an accuracy of 0.390% (1 part in 256). While using a processor of 16 bits gives an accuracy of 0.0015%, (1 part in 65,536) or 260 times more accurate. This accuracy can be maintained as digital quantities are manipulated and processed very rapidly, millions of times faster than analogue signals.

In most cases, sensors and more specifically analogue sensors generally require an external power supply and some form of additional amplification or filtering of the signal in order to produce a suitable electrical signal which is capable of being measured or used. One very good way of achieving both amplification and filtering within a single circuit is to use Operational Amplifiers.

Signal Conditioning of Sensors

Op-amps can be used to provide amplification of signals when connected in either inverting or non-inverting configurations.

The very small analogue signal voltages produced by a sensor such as a few milli-volts or even pico-volts can be amplified many times over by a simple op-amp circuit to produce a much larger voltage signal of say 5v or 5mA that can then be used as an input signal to a microprocessor or analogue-to-digital based system.

Therefore, to provide any useful signal a sensors output signal has to be amplified with an amplifier that has a voltage gain up to 10,000 and a current gain up to 1,000,000 with the amplification of the signal being linear with the output signal being an exact reproduction of the input, just changed in amplitude.

Then amplification is part of signal conditioning. So when using analogue sensors, generally some form of amplification (Gain), impedance matching, isolation between the input and output or perhaps filtering (frequency selection) may be required before the signal can be used and this is conveniently performed by Operational Amplifiers.

Also, when measuring very small physical changes the output signal of a sensor can become "contaminated" with unwanted signals or voltages that prevent the actual signal required from being measured correctly. These unwanted signals are called "Noise". This Noise or Interference can be either greatly reduced or even eliminated by using signal conditioning or filtering techniques.

By using either a Low Pass, or a High Pass or even Band Pass filter the "bandwidth" of the noise can be reduced to leave just the output signal required. For example, many types of inputs from switches, keyboards or manual controls are not capable of changing state rapidly and so low-pass filter can be used. When the interference is at a particular frequency, for example mains frequency, narrow band reject or Notch filters can be used to produce frequency selective filters.

Typical Op-amp Filters

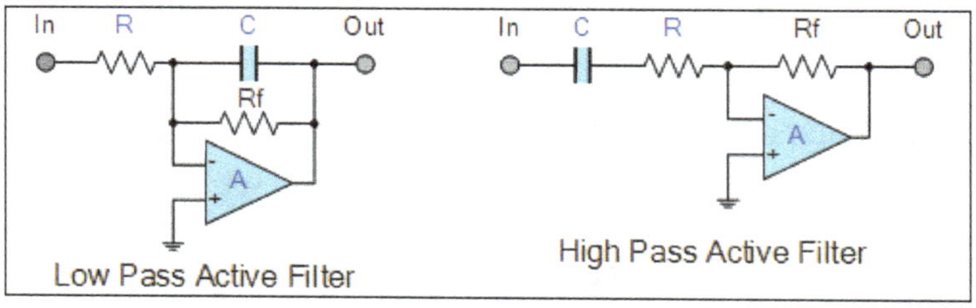

Where some random noise still remains after filtering it may be necessary to take several samples and then average them to give the final value so increasing the signal-to-noise ratio.

Actuator

An actuator is a device that uses a form of power to convert a control signal into mechanical motion. From electric door locks in automobiles, to ailerons on aircraft, actuators are all around us. Industrial plants use actuators to operate valves, dampers, fluid couplings, and other devices used in industrial process control. The industrial actuator can use air, hydraulic fluid, or electricity for motive power. These are referred to as pneumatic, electro-hydraulic, or electric actuators.

Pneumatic Actuators

Pneumatic actuators can utilize a cylinder, diaphragm, or rotary vane design. Compressed air acts upon a piston or paddle to move it in one direction. A counterforce, consisting of a spring or a second volume of compressed air, is applied to the opposite side of the piston or paddle to reverse or maintain position. For industrial control, a pneumatic actuator requires positioning electronics, and an instrument air system. Of the three types of actuators, pneumatics tend to have the lowest initial price. However, operating costs are high due to the need to generate clean, dry, compressed air. Maintenance requirements are also the highest of the three actuator types.

Hydraulic Actuator

Electro-hydraulic actuators operate a piston with pressurized oil. A motorized pump sends fluid from a reservoir through a control valve(s) to opposite sides of a cylinder. The high power density of this system allows for high thrust, and fast operating speeds. Electro-hydraulics tend to have a high initial price. Routine maintenance involves periodic replacement of seals, o-rings, etc. Since the oil used in hydraulic equipment can pose a fire hazard, this type of actuator may not be suitable for some industrial environments.

Electric actuators utilize an electric motor and gear reduction to produce a force or torque. A variety of technologies can be used for an electric actuator design. Motors may use AC or DC power, and an asynchronous (squirrel cage) or synchronous design. Gearing systems can include worm, spur, scotch yoke, among others. Gear lubrication may consist of an oil filled gear box, or heavy duty grease applied to the gear surfaces. Various accessories are often available to supervise and report on the actuator's status and operating conditions. There is a wide variety of electric actuators on the market with different technologies, prices, performance, and quality.

Because actuators are closely integrated with driven elements like valves, and dampers, they are often referred to as assemblies. Examples include motor operated valves (MOV), air operated valves (AOV), and damper drives. Frequently, the actuator may be assumed to come with the equipment package. Examples include control valve, or air damper. In practice, actuators are the critical link between the control system and the final element. Its performance can significantly impact many aspects of an industrial plant's operations and products.

When selecting an actuator, the actuator should have performance characteristics that will enable a control system to perform as designed.

The key actuator performance characteristics are as follows:

- Precise, repeatable positioning typically better than 0.15% of span.
- The ability to start and stop instantaneously without dead time or position overshoot.
- Continuous duty rating without limitations on the number of starts per minute.
- Perform consistently and unaffected by load.
- Rugged industrial design capable of operating in difficult environments without an effect on performance.
- Minimal periodic maintenance required.

An actuator designed with these characteristics provides two extremely important advantages:

- An ability to follow the demand signal from the controller precisely and instantly. This ensures that the actuator responds exactly as directed by the controller. Thus, the actuator is not the limiting factor in the control loop and the controller can function to its optimal levels.
- A high degree of maintenance-free reliability. An actuator designed to function as outlined above by default is more rugged than typical actuators. By design, then, it is capable of a much higher degree of reliability.

Transducers 2

- **Resistive Transducer**
- **Capacitive Transducer**
- **Piezoelectric Transducer**
- **Accelerometer**
- **Pressure Transducer**
- **Capacitive Micromachined Ultrasonic Transducer**
- **Electromagnetic Acoustic Transducer**
- **Orthomode Transducer**

Transducers are the devices that convert energy from one form to another. There are various types of transducers such as resistive transducers, capacitive transducers, piezoelectric transducers, pressure transducers, capacitive micromachined ultrasonic transducers, electromagnetic acoustic transducers and orthomode transducers. All these types of transducers have been carefully analyzed in this chapter.

Transducer is a device that converts input energy into output energy, the latter usually differing in kind but bearing a known relation to input. Originally, the term referred to a device that converted mechanical stimuli into electrical output, but it has been broadened to include devices that sense all forms of stimuli—such as heat, radiation, sound, strain, vibration, pressure, acceleration, and so on—and that can produce output signals other than electrical—such as pneumatic or hydraulic. Many measuring and sensing devices, as well as loudspeakers, thermocouples, microphones, and phonograph pickups, may be termed transducers.

There are hundreds of kinds of transducers, many of which are designated by the energy change they accomplish. For example, piezoelectric transducers contain a piezoelectric element that produces motion when subjected to an electrical voltage or produces electrical signals when subjected to strain. The latter effect may be applied in an accelerometer, a piezoelectric vibration pickup, or a strain gauge. An electroacoustic transducer may convert electrical signals to acoustic signals or vice versa. An example is the hydrophone, which responds to waterborne sound waves and is useful in underwater sound detection. A photoelectric transducer reacts to visible light to produce electrical energy. Electromagnetic transducers form a large group, the major categories of which are differential transformers, Hall-effect magnetic transducers, inductance transducers, induction transducers, and saturable reactors. These operate on electromagnetic principles.

Electrical transducers may be classified as active or passive. The active transducers generate electric current or voltage directly in response to stimulation. An example is the thermocouple; here, the fact that a current will flow in a continuous circuit of two metals, if the two junctions are at different temperatures, is used to generate electricity. The passive transducer produces a change in some passive electrical quantity, such as capacitance, resistance, or inductance, as a result of stimulation. Passive transducers usually require additional electrical energy. A simple example of a passive transducer is a device containing a length of wire and a moving contact touching the wire. The position of the contact determines the effective length of the wire and, thus, the resistance offered to electric current flowing through it. This is the simplest version of what is called a linear-displacement transducer, or linear potentiometer. For practical use, such transducers employ wire-wound, thin-film, or printed circuits to allow for a long resistor within a relatively small device. The longer the resistor, the greater the drop in voltage passing through the device; thus, changes in position are converted to electrical signals.

Transducers also may produce pneumatic or hydraulic output. Pneumatic systems communicate by means of compressed air. An example is a device in which motion is applied via a system of pivots to a baffle that can be moved closer to or farther away from a nozzle that emits a stream of air. The amount of resistance created by the baffle affects the amount of back pressure behind the nozzle, creating a pneumatic signal. Hydraulic systems tend to be designed similarly to pneumatic systems, except that hydraulic systems use hydraulic (liquid) pressure rather than air pressure. Fluidic principles, which apply to the interaction between two fluid streams, have also been used to create transducers.

Types of Transducers

The transducer is of many types, and they can be classified by the following criteria:

1. By transduction used.
2. As a primary and secondary transducer.
3. As a passive and active transducer.
4. As analogue and digital transducer.
5. As the transducer and inverse transducer.

The transducer receives the measure and and gives a proportional amount of output signal. The output signal is sent to the conditioning device where the signal is attenuated, filtered, and modulated.

The input quantity is the non-electrical quantity, and the output electrical signal is in the form of the current, voltage or frequency.

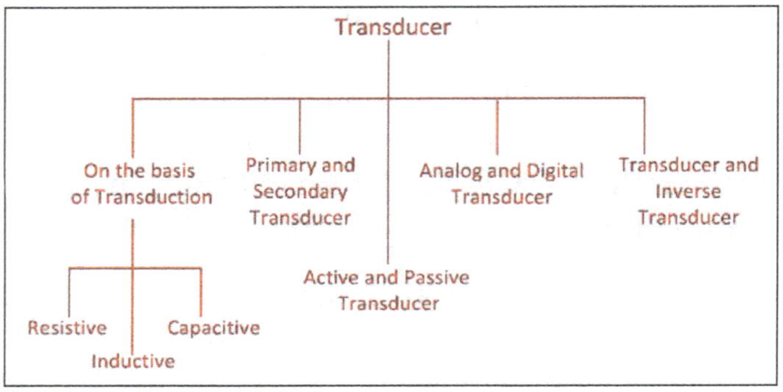

Classification based on the Principle of Transduction

The transducer is classified by the transduction medium. The transduction medium may be resistive, inductive or capacitive depends on the conversion process that how input transducer converts the input signal into resistance, inductance and capacitance respectively.

Primary and Secondary Transducer

- Primary Transducer – The transducer consists the mechanical as well as the electrical devices. The mechanical devices of the transducer change the physical input quantities into a mechanical signal. This mechanical device is known as the primary transducers.
- Secondary Transducer – The secondary transducer converts the mechanical signal into an electrical signal. The magnitude of the output signal depends on the input mechanical signal.

Example of Primary and Secondary Transducer

Consider the Bourdon's Tube shown in the figure below. The tube act as a primary transducer. It detects the pressure and converts it into a displacement from its free end. The displacement of the free ends moves the core of the linear variable displacement transformer. The movement of the core induces the output voltage which is directly proportional to the displacement of the tube free end.

Thus, the two type of transduction occurs in the Bourdon's tube. First, the pressure is converted into a displacement and then it is converted into the voltage by the help of the L.V.D.T. The Bourdon's Tube is the primary transducer, and the L.V.D.T is called the secondary transducer.

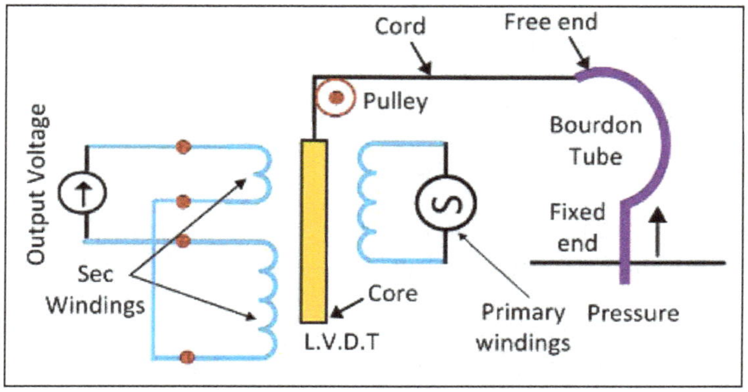

Bourdon's Tube

Passive and Active Transducer

The transducer is classified as the active and passive transducer:

- Passive Transducer – The transducer which requires the power from an external supply source is known as the passive transducer. They are also known as the external power transducer. The capacitive, resistive and inductive transducers are the example of the passive transducer.
- Active Transducer – The transducer which does not require the external power source is known as the active transducer. Such type of transducer develops theirs owns voltage or current, hence known as a self-generating transducer. The output signal is obtained from the physical input quantity.

The physical quantity like velocity, temperature, force and the intensity of light is induced with the help of the transducer. The piezoelectric crystal, photo-voltaic cell, tacho generator, thermocouples, photovoltaic cell are the examples of the active transducers.

Examples: Consider the examples of a piezoelectric crystal. The crystal is sandwiched between the two metallic electrodes, and the entire sandwiched is fastened to the base. The mass is placed on the top of the sandwiched.

The piezo crystal has the special property because of which when the force is applied to the crystal, they induce the voltage. The base provides the acceleration due to which the voltage is generated. The mass applies on the crystals induces an output voltage. The output voltage is proportional to the acceleration.

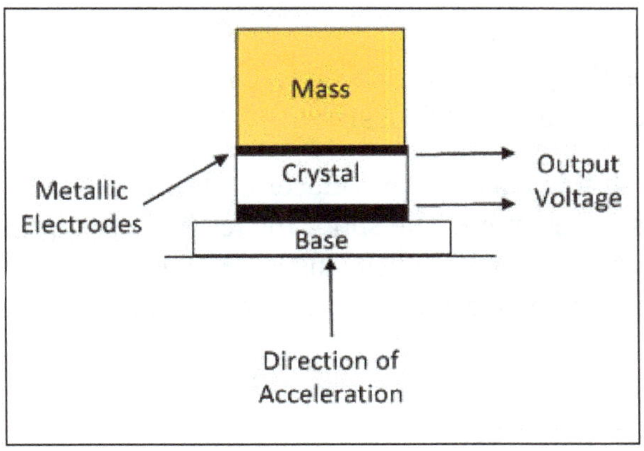

The above mentioned transducer is known as the accelerometer which converts the acceleration into an electric voltage. This transducer does not require any auxiliary power source for the conversion of physical quantity into an electrical signal.

Analog and Digital Transducer

The transducer can also be classified by their output signals. The output signal of the transducer may be continuous or discrete.

Analog Transducer – The Analog transducer changes the input quantity into a continuous function. The strain gauge, L.V.D.T, thermocouple, thermistor are the examples of the analogue transducer.

Digital Transducer – These transducers convert an input quantity into a digital signal or in the form of the pulse. The digital signals work on high or low power.

Transducer and Inverse Transducer

- Transducer – The device which converts the non-electrical quantity into an electric quantity is known as the transducer.
- Inverse Transducer – The transducer which converts the electric quantity into a physical quantity, such type of transducers is known as the inverse transducer. The transducer has high electrical input and low non-electrical output.

Resistive Transducer

The transducer whose resistance varies because of the environmental effects such type of transducer is known as the resistive transducer. The change in resistance is measured by the ac or dc measuring devices. The resistive transducer is used for measuring the physical quantities like temperature, displacement, vibration etc.

The measurement of the physical quantity is quite difficult. The resistive transducer converts the physical quantities into variable resistance which is easily measured by the meters. The process of variation in resistance is widely used in the industrial applications.

The resistive transducer can work both as the primary as well as the secondary transducer. The primary transducer changes the physical quantities into a mechanical signal, and secondary transducer directly transforms it into an electrical signal.

Example: The circuit of the sliding resistive transducer is shown in the figure below. The sliding contacts are placed on the resistive element. The slider moves horizontally. The movement of the slider changes the value of the resistive element of the transducer which is measured by the voltage source E. The displacement of the slider is converted into an electrical signal.

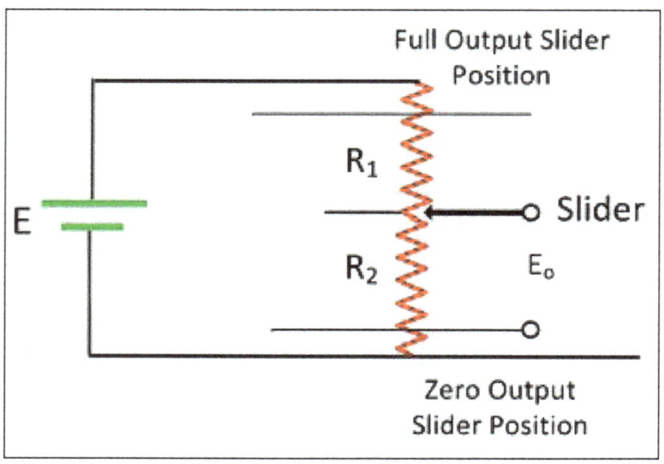

Sliding Resistive Transducer

Advantages of Resistive Transducer

The following are the advantages of the resistive transducer:

1. Both the AC and DC, current or voltage is appropriate for the measurement of variable resistance.
2. The resistive transducer gives the fast response.
3. It is available in various sizes and having a high range of resistance.

Working Principle of Resistive Transducer

The resistive transducer element works on the principle that the resistance of the element is directly proportional to the length of the conductor and inversely proportional to the area of the conductor.

$$R = \rho L / A$$

where,

- R – resistance in ohms.
- A – cross-section area of the conductor in meter square.
- L – Length of the conductor in meter square.
- ρ– the resistivity of the conductor in materials in ohm meter.

The resistive transducer is designed by considering the variation of the length, area and resistivity of the metal.

Applications of Resistive Transducer

The following are the applications of the resistive transducer.

- Potentiometer – The translation and rotatory potentiometer are the examples of the resistive transducers. The resistance of their conductor varies with the variation in their lengths which is used for the measurement of displacement.
- Strain gauges – The resistance of their semiconductor material changes when the strain occurs on it. This property of metals is used for the measurement of the pressure, force-displacement etc.
- Resistance Thermometer – The resistance of the metals changes because of changes in temperature. This property of conductor is used for measuring the temperature.
- Thermistor – It works on the principle that the temperature coefficient of the thermistor material varies with the temperature. The thermistor has the negative temperature coefficient. The Negative temperature coefficient means the temperature is inversely proportional to resistance.

There are a number of ways because of which the resistance of the metal changes with the changed in the physical phenomenon. And this property of conductors is used for measuring the physical quantities of material.

Capacitive Transducer

The capacitive transducer is used for measuring the displacement, pressure and other physical quantities. It is a passive transducer that means it requires external power for operation. The capacitive transducer works on the principle of variable capacitances. The capacitance of the capacitive transducer changes because of many reasons like overlapping of plates, change in distance between the plates and dielectric constant.

The capacitive transducer contains two parallel metal plates. These plates are separated by the dielectric medium which is either air, material, gas or liquid. In the normal capacitor the distance between the plates are fixed, but in capacitive transducer the distance between them are varied.

The capacitive transducer uses the electrical quantity of capacitance for converting the mechanical movement into an electrical signal. The input quantity causes the change of the capacitance which is directly measured by the capacitive transducer.

The capacitors measure both the static and dynamic changes. The displacement is also measured directly by connecting the measurable devices to the movable plate of the capacitor. It works on with both the contacting and non-contacting modes.

Principle of Operation

The equations below express the capacitance between the plates of a capacitor:

$$C = \varepsilon A / d$$

$$C = \varepsilon_r \varepsilon_0 A / d$$

where,

- A – overlapping area of plates in m^2.
- d – the distance between two plates in meter.
- ε – permittivity of the medium in F/m.
- ε_r – relative permittivity.
- ε_0 – the permittivity of free space.

The schematic diagram of a parallel plate capacitive transducer is shown in the figure below.

The change in capacitance occurs because of the physicals variables like displacement, force, pressure, etc. The capacitance of the transducer also changes by the variation in their dielectric constant which is usually because of the measurement of liquid or gas level.

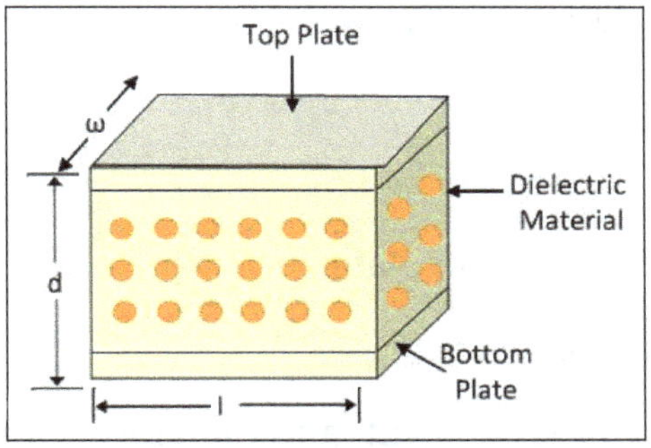

Parallel Plate Capacitive Transducer

The capacitance of the transducer is measured with the bridge circuit. The output impedance of transducer is given as:

$$X_C = 1/2\pi f c$$

where, C – capacitance.

f – frequency of excitation in Hz.

The capacitive transducer is mainly used for measurement of linear displacement. The capacitive transducer uses the following three effects.

1. Variation in capacitance of transducer is because of the overlapping of capacitor plates.
2. The change in capacitance is because of the change in distances between the plates.
3. The capacitance changes because of dielectric constant.

The following methods are used for the measuring displacement:

1. A transducer using the change in the Area of Plates – The equation below shows that the capacitance is directly proportional to the area of the plates. The capacitance changes correspondingly with the change in the position of the plates. The capacitive transducers are used for measuring the large displacement approximately from 1mm to several cms. The area of the capacitive transducer changes linearly with the capacitance and the displacement. Initially, the nonlinearity occurs in the system because of the edges. Otherwise, it gives the linear response.

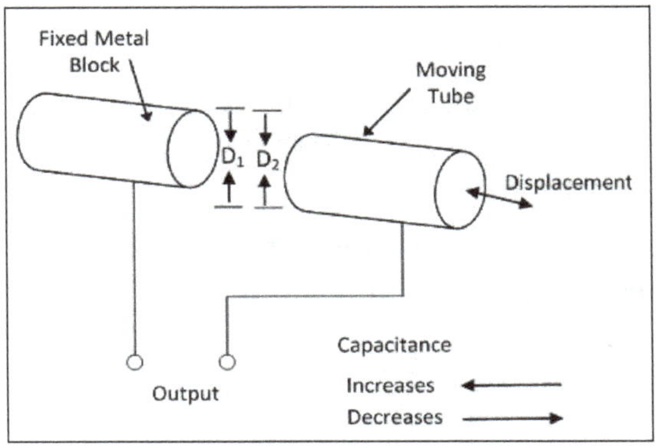

Capacitive Tranducer

The capacitance of the parallel plates is given as:

$$C = \frac{\varepsilon A}{d} = \frac{\varepsilon x \omega}{d} F$$

where, x – the length of overlapping part of plates.

ω – the width of overlapping part of plates.

The sensitivity of the displacement is constant, and therefore it gives the linear relation between the capacitance and displacement:

$$S = \frac{\partial C}{\partial x} = \varepsilon \frac{\omega}{d} F / m$$

The capacitive transducer is used for measuring the angular displacement. It is measured by the movable plates shown below. One of the plates of the transducer is fixed, and the other is movable.

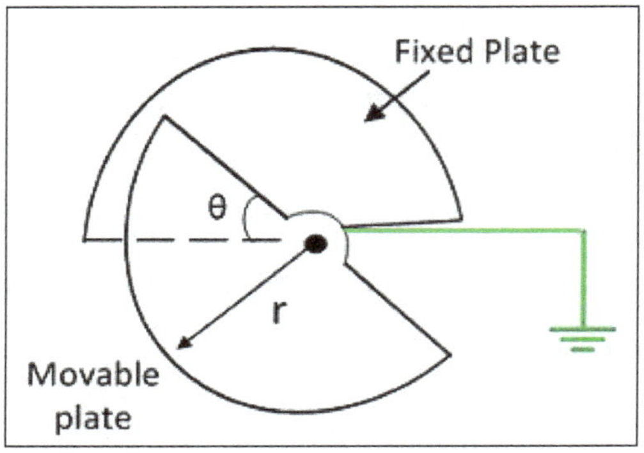

The phasor diagram of the transducer is shown in the figure below.

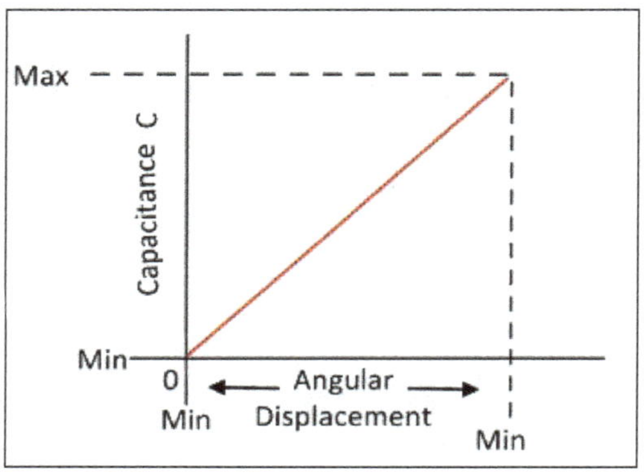

The angular movement changes the capacitance of the transducers. The capacitance between them is maximum when these plates overlap each other. The maximum value of capacitance is expressed as:

$$C_{\max} = \frac{\varepsilon C}{d} = \frac{\pi \varepsilon r^2}{2d}$$

The capacitance at angle θ is given expressed as:

$$C = \frac{\varepsilon \theta r^2}{2d}$$

θ – angular displacement in radian. The sensitivity for the change in capacitance is given as:

$$S = \frac{\partial C}{\partial \theta} = \frac{\varepsilon r^2}{2d}$$

The 180° is the maximum value of the angular displacement of the capacitor.

2. The transducer using the change in distance between the plates – The capacitance of the transducer is inversely proportional to the distance between the plates. The one plate of the transducer is fixed, and the other is movable. The displacement which is to be measured links to the movable plates.

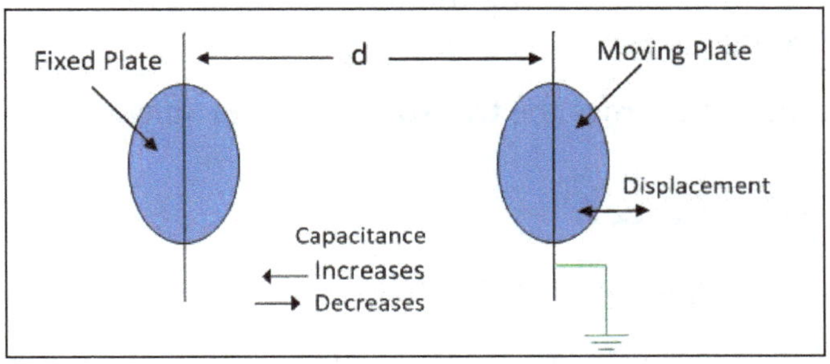

The capacitance is inversely proportional to the distance because of which the capacitor shows the nonlinear response. Such type of transducer is used for measuring the small displacement. The phasor diagram of the capacitor is shown in the figure below.

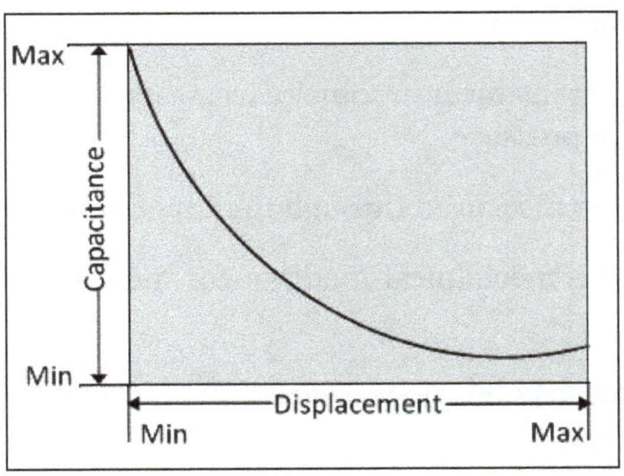

The sensitivity of the transducer is not constant and vary from places to places.

Advantages of Capacitive Transducer

The following are the major advantages of capacitive transducers:

1. It requires an external force for operation and hence very useful for small systems.

2. The capacitive transducer is very sensitive.
3. It gives good frequency response because of which it is used for the dynamic study.
4. The transducer has high input impedance hence they have a small loading effect.
5. It requires small output power for operation.

Disadvantages of Capacitive Transducer

The main disadvantages of the transducer are as follows:

1. The metallic parts of the transducers require insulation.
2. The frame of the capacitor requires earthing for reducing the effect of the stray magnetic field.
3. Sometimes the transducer shows the nonlinear behaviours because of the edge effect which is controlled by using the guard ring.
4. The cable connecting across the transducer causes an error.

Uses of Capacitive Transducer

The following are the uses of capacitive transducer:

1. The capacitive transducer uses for measurement of both the linear and angular displacement. It is extremely sensitive and used for the measurement of very small distance.
2. It is used for the measurement of the force and pressures. The force or pressure, which is to be measured is first converted into a displacement, and then the displacement changes the capacitances of the transducer.
3. It is used as a pressure transducer in some cases, where the dielectric constant of the transducer changes with the pressure.
4. The humidity in gases is measured through the capacitive transducer.
5. The transducer uses the mechanical modifier for measuring the volume, density, weight etc.

The accuracy of the transducer depends on the variation of temperature to the high level.

Piezoelectric Transducer

A piezoelectric crystal transducer/sensor is an active sensor and it does not need the help of an external power as it is self-generating. It is important to know the basics of a piezoelectric quartz crystal and piezoelectric effect before going into details about the transducer.

Piezoelectric Quartz Crystal

A quartz crystal is a piezoelectric material that can generate a voltage proportional to the stress applied upon it. For the application, a natural quartz crystal has to be cut in the shape of a thin plate of rectangular or oval shape of uniform thickness. Each crystal has three sets of axes – Optical axes, three electrical axes OX1, OX2, and OX3 with 120 degree with each other, and three mechanical axes OY1,OY2 and OY3 also at 120 degree with each other. The mechanical axes will be at right angles to the electrical axes. Some of the parameters that decide the nature of the crystal for the application are:

1. Angle at which the wafer is cut from natural quartz crystal.
2. Plate thickness.
3. Dimension of the plate.
4. Means of mounting.

Piezoelectric Effect

The X-Y axis of a piezoelectric crystal and its cutting technique is shown in the figure below.

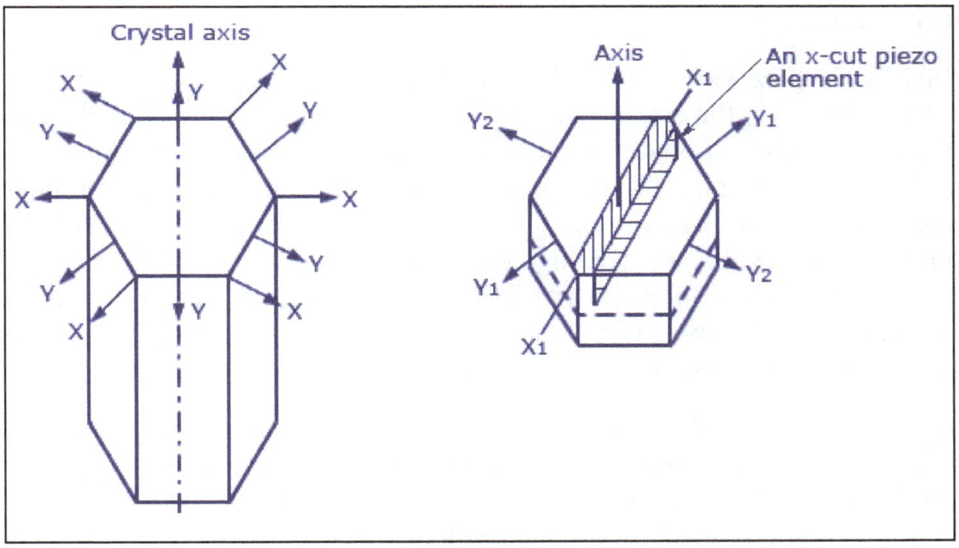

X-Y Axes of a Piezoelectric Crystal

The direction, perpendicular to the largest face, is the cut axis referred to. If an electric stress is applied in the directions of an electric axis (X-axis), a mechanical strain is produced in the direction of the Y-axis, which is perpendicular to the relevant X-axis. Similarly, if a mechanical strain is given along the Y-axis, electrical charges will be produced on the faces of the crystal, perpendicular to the X-axis which is at right angles to the Y-axis.

Some of the materials that inherit piezo-electric effect are quartz crystal, Rochelle salt, barium titanate, and so on. The main advantages of these crystals are that they have high mechanical and thermal state capability, capability of withstanding high order of strain, low leakage, and good frequency response, and so on.

A piezoelectric transducer may be operated in one of the several modes as shown in the figure below.

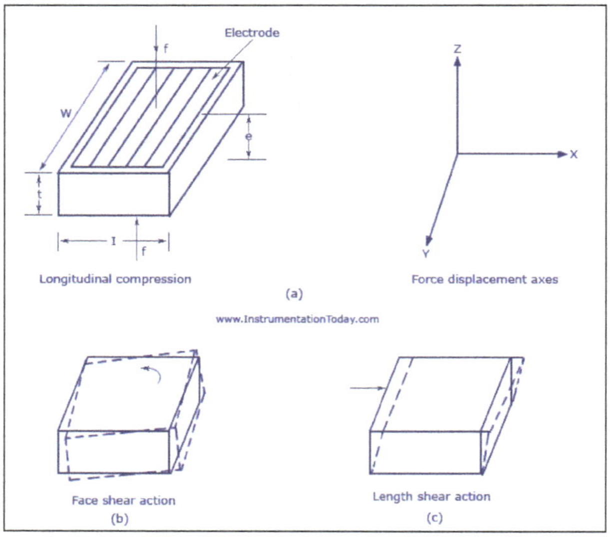

Piezoelectric Crystal

Piezoelectric Transducer

The main principle of a piezoelectric transducer is that a force, when applied on the quartz crystal, produces electric charges on the crystal surface. The charge thus produced can be called as piezo-electricity. Piezo electricity can be defined as the electrical polarization produced by mechanical strain on certain class of crystals. The rate of charge produced will be proportional to the rate of change of force applied as input. As the charge produced is very small, a charge amplifier is needed so as to produce an output voltage big enough to be measured. The device is also known to be mechanically stiff. For example, if a force of 15 kiloN is given to the transducer, it may only deflect to a maximum of 0.002mm. But the output response may be as high as 100KiloHz. This proves that the device is best applicable for dynamic measurement.

The figure shows a conventional piezoelectric transducer with a piezoelectric crystal inserted between a solid base and the force summing member. If a force is applied on the pressure port, the same force will fall on the force summing member. Thus a potential difference will be generated on the crystal due to its property. The voltage produced will be proportional to the magnitude of the applied force.

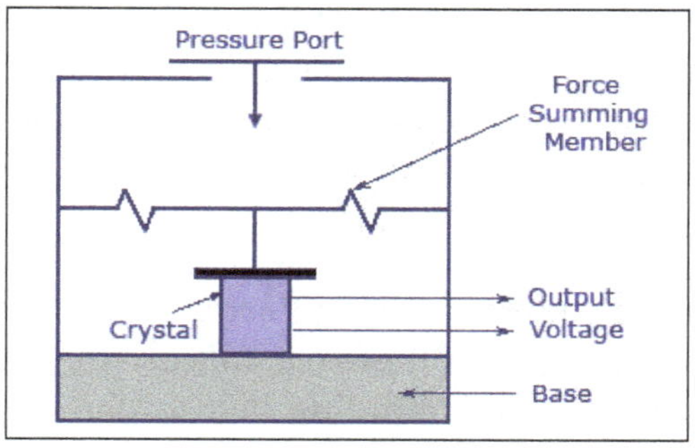

Piezoelectric Transducer

Piezoelectric Transducer can measure pressure in the same way a force or an acceleration can be measured. For low pressure measurement, possible vibration of the amount should be compensated for. The pressure measuring quartz disc stack faces the pressure through a diaphragm and on the other side of this stack, the compensating mass followed by a compensating quartz.

Applications

1. Due to its excellent frequency response, it is normally used as an accelerometer, where the output is in the order of (1-30) mV per gravity of acceleration.
2. The device is usually designed for use as a pre-tensional bolt so that both tensional and compression force measurements can be made.
3. Can be used for measuring force, pressure and displacement in terms of voltage.

Advantages

1. Very high frequency response.
2. Self generating, so no need of external source.
3. Simple to use as they have small dimensions and large measuring range.
4. Barium titanate and quartz can be made in any desired shape and form. It also has a large dielectric constant. The crystal axis is selectable by orienting the direction of orientation.

Disadvantages

1. It is not suitable for measurement in static condition.
2. Since the device operates with the small electric charge, they need high impedance cable for electrical interface.
3. The output may vary according to the temperature variation of the crystal.
4. The relative humidity rises above 85% or falls below 35%, its output will be affected. If so, it has to be coated with wax or polymer material.

Accelerometer

An accelerometer is a transducer that is used to measure the physical or measurable acceleration that is made by an object. Co-ordinate acceleration cannot be measured using this device.

An accelerometer is an electro-mechanical device that is used to measure the specific force of an object, a force obtained due to the phenomenon of weight exerted by an object that is kept in the frame of reference of the accelerometer.

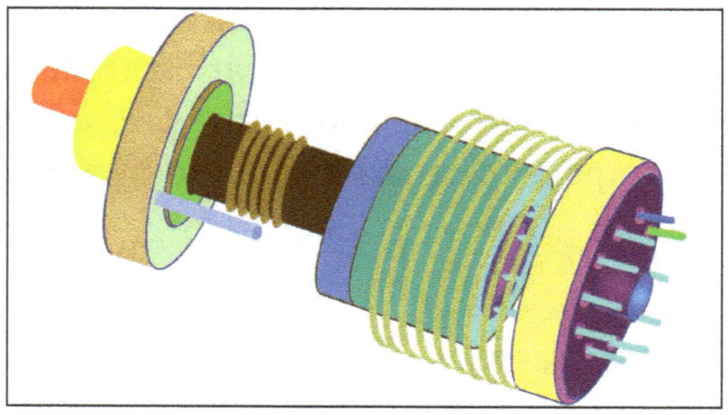

Accelerometer

In the case of static acceleration, the device is mainly used to find the degrees at which an object is tilted with respect to the ground. In dynamic acceleration, the movement of the object can be foreseen.

The most commonly used device is the piezoelectric accelerometer. As the name suggests, it uses the principle of piezoelectric effect. The device consists of a piezoelectric quartz crystal on which an accelerative force, whose value is to be measured, is applied.

Due to the special self-generating property, the crystal produces a voltage that is proportional to the accelerative force. The working and the basic arrangement is shown in the figure.

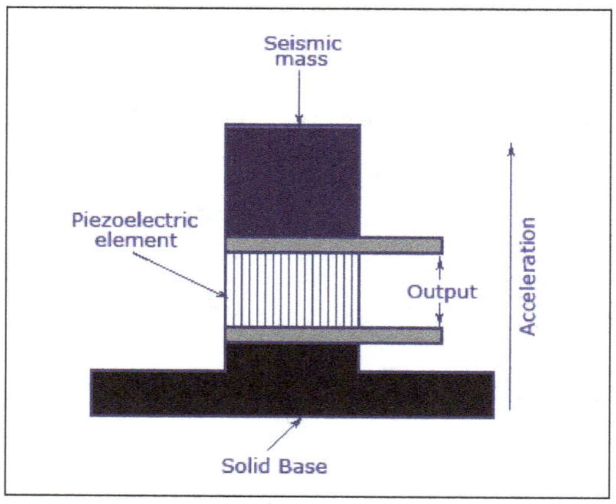

Piezoelectric Accelerometer

As the device finds its application as a highly accurate vibration measuring device, it is also called a vibrating sensor. Vibration sensors are used for the measurement of vibration in bearings of heavy equipment and pressure lines. The piezoelectric accelerometer can be classified into two. They are high impedance output accelerometer and low impedance accelerometer.

In the case of high impedance device, the output voltage generated in proportion to the acceleration is given as the input to other measuring instruments. The instrumentation process and signal conditioning of the output is considered high and thus a low impedance device cannot be of any use for this application. This device is used at temperatures greater than 120 degree Celsius.

The low impedance accelerometer produces a current due to the output voltage generated and this charge is given to a FET transistor so as to convert the charge into a low impedance output. This type of accelerometers is most commonly used in industrial applications.

Types

In industrial applications, the most commonly used components to convert the mechanical action into its corresponding electrical output signal are piezoelectric, piezoresistive and capacitive in nature. Piezoelectric devices are more preferred in cases where it is to be used in very high temperatures, easy mounting and also high frequency rang e up to 100 kilohertz. Piezoresistive devices are used in sudden and extreme vibrating applications. Capacitive accelerometers are preferred in applications such as a silicon-micro machined sensor material and can operate in frequencies up to 1 kilohertz. All these devices are known to have very high stability and linearity.

Nowadays, a new type of accelerometer called the Micro Electro-Mechanical System (MEMS) Accelerometer is being used as it is simple, reliable and highly cost effective. It consists of a cantilever beam along with a seismic mass which deflects due to an applied acceleration. This deflection is measured using analog or digital techniques and will be a measure of the acceleration applied.

Accelerometer Specifications

1. Frequency Response – This parameter can be found out by analyzing the properties of the quartz crystal used and also the resonance frequency of the device.
2. Accelerometer Grounding – Grounding can be in two modes. One is called the Case Grounded Accelerometer which has the low side of the signal connected to their core. This device is susceptible to ground noise. Ground Isolation Accelerometer refers to the electrical device kept away from the case. Such a device is prone to ground produced noise.
3. Resonant Frequency – It should be noted that the resonant frequency should be always higher the the frequency response.
4. Temperature of Operation – An accelerometer has a temperature range between -50 degree Celsius to 120 degree Celsius. This range can be obtained only by accurate installment of the device.
5. Sensitivity – The device must be designed in such a way that it has higher sensitivity. That is, even for a small accelerative force, the electrical output signal should be very high. Thus a high signal can be measured easily and is sure to be accurate.
6. Axis – Most of the industrial applications requires only a 2-axis accelerometer. But if you want to go for 3D positioning, a 3-axis accelerometer will be needed.
7. Analog/Digital Output – You must take special care in choosing the type of output for the device. Analog output will be in the form of small changing voltages and digital output will be in PWM mode.

Device Selection

Selection of the device depends on the following factors.

1. Selection depends on the range of frequency you need.
2. Depends on the size and shape of the object whose acceleration is to be measured.

3. Whether the measurement environment is dirty or clean.
4. Depends on the range of vibration that is to be measured.
5. Depends on the range of temperature in which the device will have to work.
6. Depends on whether the device is to be case grounded or grounded isolated.

Applications

1. Machine monitoring.
2. Used to measure earthquake activity and aftershocks.
3. Used in measuring the depth of CPR chest compression.
4. Used in Internal Navigation System (INS). That is, measuring the position, orientation, and velocity of an object in motion without the use of any external reference.
5. Used in airbag shooting in cars and vehicle stability control.
6. Used in video games like PlayStation 3, so as to make the steering more controlled, natural and real.
7. Used in camcorder to make images stable.

Pressure Transducer

A pressure transducer is used to convert a certain value of pressure into its corresponding mechanical or electrical output. Measurement if pressure is of considerable importance in process industries.

Types

The types of pressure sensors are differentiated according to the amount of differential pressure they are able to measure.

For low differential pressure measurement Liquid Column Manometers are used. Elastic type pressure gauges are also used for pressure measurement up to 700 MPa. Some of the common elastic/mechanical types are:

- Bourdon Tubes
- Diaphragm
- Piston Type Pressure Transducer
- Bellows

Other Types:

- Electrical Type Pressure Sensors
- Bell Gauge
- Manometers

Before going into further details regarding pressure measurement, it is important to know the different terms related to pressure.

Basic Terms Related to Pressure Measurement

Pressure

Pressure is known to be the force that is exerted due to the weights of different gases and liquids. Some common examples are atmospheric pressure and the pressure implied by liquids inside the walls and underside of a container.

Pressure can be measured as the force exerted over a certain area.

Pressure = Force/Area

Pressure is not an independent variable as it is derived from force and area and it is not ideal as it depends on other factors like elevation, fluid density, temperature, flow velocity, and so on.

In instrumentation analysis, pressure is commonly expressed in pounds per square inch (psi). It can also be expressed in pounds per square feet (psf) and Pascals (Pa). Pascal is the SI unit if pressure. In many cases, pressure is expressed in terms of atmosphere which is the height of the barometric column at zero degrees Celsius, being equal to 76 cm of mercury or equivalent to 14.696 pounds per square inch absolute, 1 kg/cm^2. Most of the pressures range from a little below atmosphere to hundreds of atmospheres.

- Density is the mass per unit vlume of the material. It can be expressed as kilogram per cubic meter (kg/m^3).
- Specific Weight is the weight per unit volume of the material. It can be expressed as Newon per cubic meter (N/m^3).
- Specific Gravity is basically a non-dimensional value as it is the ratio of two measurements in the same unit. It can be the ratio between the density of a material and the density of water or even the ratio between the specific weight of a material to the specific weight of water.
- Static Pressure is the fluid or gas pressure that does not move.
- Dynamic Pressure is the gas or fluid pressure that is obtained when it impacts with a surface or an object due to its motion or flow.
- Impact Pressure is the total pressure or the addition of both the static and dynamic pressures.
- Atmospheric Pressure is the surface pressure of earth and is available due to the weight if the gases in the earth's atmosphere.

- Another important aspect of pressure measurement is the measurement of very low pressure or what is known as vacuum. With the advancement of scientific research and industrial application of the results, pressure is as low as 10-6 mm of mercury is often required to be measured in some systems. Measurement of pressure, therefore, consists of two parts – that of pressure and vacuum. The force exerted by the fluid per unit area of the wall of the container is called the absolute pressure, whereas the gauge pressure is the difference between the absolute and local atmospheric pressure, and when gauge pressure is negative, it is known as vacuum.

Basic methods of pressure measurement are same as those of force measurement. For high vacuum, however, some special techniques are necessary. Primary sensors are mostly, mechanical which through secondary sensing means provide electrical outputs. Manometers and elastic element sensors are used as primary pressure sensors while secondary sensing, often called transducing here, involves resistive, inductive and capacitive changes for deriving electrical outputs.

Pressure Instrument Selection

Selection of pressure instrument for a particular application must be done carefully, taking into consideration various aspects such as process conditions, turn down requirements, accuracy, installation requirements, and so on.

While selecting a pressure instrument for a particular application, the process data such as fluid phase, pressure, temperature, density and viscosity must be correctly defined for all operating conditions including start-up, emergency operations and design conditions.

Another important parameter for selection is the turn down requirements, based on which we can select the pressure instrument to suit the maximum and minimum conditions within its specified accuracy limits.

In addition to the above, the installation requirements of the selected pressure instrument shall be carefully addressed taking in to account visibility, accessibility, because these requirements may affect the piping layouts.

Piston Type Pressure Transducer

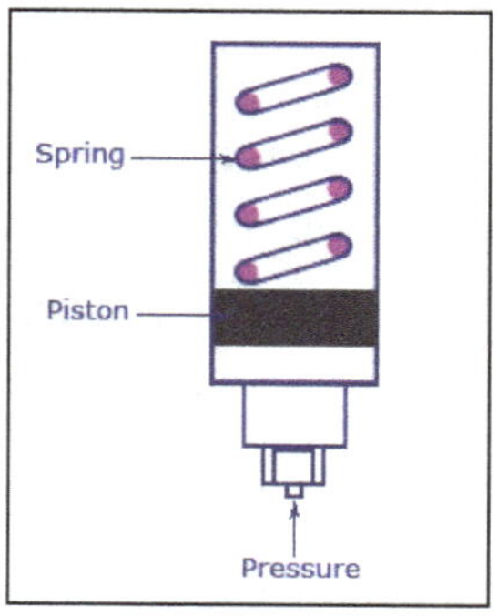

Piston Type Pressure Transducer.

As shown in the figure below, the input pressure is given to the piston. This moves the piston accordingly and causes the spring to be compressed. The piston position will be directly proportional to the amount of input pressure exerted. A meter is placed outside the piston and spring arrangement, which indicates the amount of pressure exerted. As the device has the ability to withstand shock, sudden pressure changes, and vibrations, it is commonly used in hydraulic applications. Mostly, the output of the piston and spring arrangement is given to a secondary device to convert movement into an electrical signal.

Bell Gauge

The bell gauge is a type of pressure transducer that measures differential pressure between 0.06 Pa and 4 KPa. The static pressure may be as high as 4 to 6 MPa. The schematic diagram of a single element bell gauge is shown below.

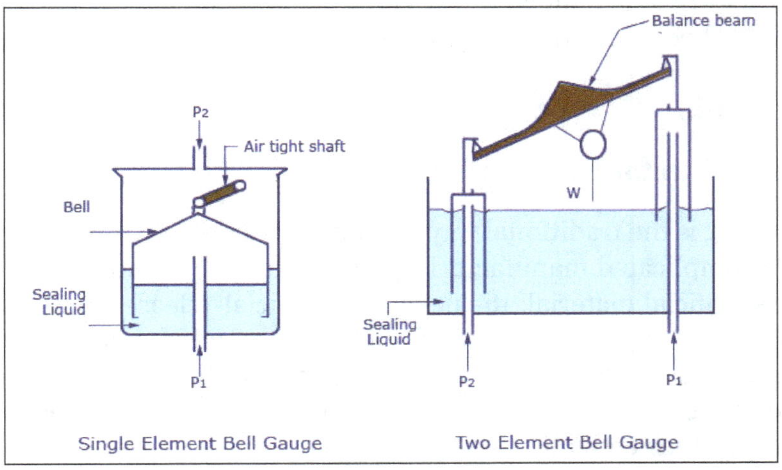

Bell Gauge.

The movement of the bell is taken out by link and lever mechanism or by some electrical methods. When the bell moves maximum up or down it closes the inlets of pressure p2 or p1, whereby protection to overrange and reversal of pressure are afforded. The two identical bells are suspended from the two knife edges of a balance beam. The differential weight is balanced statically by the movement of the counter weight w.

Capacitive Micromachined Ultrasonic Transducer

A capacitive micromachined ultrasonic transducer (CMUT) is a relatively new concept in the field of ultrasonic transducers. Most of the commercial ultrasonic transducers today are based on piezoelectricity. CMUTs are the transducers where the energy transduction is due to change in capacitance. CMUTs are constructed on silicon using micromachining techniques. A cavity is formed in a silicon substrate, and a thin layer suspended on the top of the cavity serves as a membrane on which a metallized layer acts an electrode, together with the silicon substrate which serves as a bottom electrode.

If an AC signal is applied across the biased electrodes, the vibrating membrane will produce ultrasonic waves in the medium of interest. In this way it works as a transmitter. On the other hand, if ultrasonic waves are applied on the membrane of a biased CMUT, it will generate alternating signal as the capacitance of the CMUT is varied. In this way, it works as a receiver of ultrasonic waves.

As CMUTs are micromachined devices, it is easier to construct 2D arrays of transducers using this technology. This means large numbers of CMUTs could be included in a transducer array providing larger bandwidth compared to other transducer technologies. To achieve a high frequency operation using CMUTs is easier due to its smaller dimensions. The frequency of operation depends on the cell size (cavity of membrane), and on the stiffness of the material used as a membrane. As it is built on silicon, the integration of electronics would be easier for the CMUTs compared to other transducer technologies. The properties to use in high frequency with large bandwidth makes it a good choice to use as a transducer in medical imaging, especially in an intravascular ultrasound (IVUS). Because of its broader bandwidth, it could be used in second-harmonic imaging. Also some experiments have been performed to use CMUTs as hydrophones.

Fabrication Methods

Sacrificial Release Surface Micromachining

Surface micromachining is the traditional way of manufacturing CMUTs. The major limitations of this method include complicated manufacturing process for constructing and sealing etch/drainage channels of the sacrificial material; the need for sacrificial-release channels reduces the available space for transducers, thereby reducing the achievable sound generation capability; limited control of layers' thickness during the manufacturing process; limited cavity thickness due to residues of fluid inside the cell cavity, which can cause stiction between the upper and lower parts of the cell, if the cell is not thick enough.

Wafer Bonding

Wafer bonding is the most popular method. In this method, a CMUT is built from two separate wafers, which are later bonded to achieve cells with cavities.

Fusion-bonding

Multi-user MUMPS (polyMUMPS) process. CMUTs manufactured in the multi-user MUMPS were reported to have reduced performance, such as relatively low resonating frequency.

Anodic Bonding

In anodic bonding, wafers are sealed at high temperature and in the presence of electric field.

Top-down Process

In this method the manufacturing is performed in reverse order, compared to the traditional way. The structural membrane is in silicon-nitride LPCVD, but the entire process is low-temperature, so it is CMOS-compatible. There are no etch-hole on the radiating surface of the device. The connection pads are on the back of the device, without using of through VIAs in the silicon, and the silicon substrate is completely removed. A custom acoustic backing is used to improve acoustic performances of the device. The process uses few masks.

Integration with Electrical Circuits

As mentioned earlier, one of the significant advantages of CMUTs over piezoelectric transducers is the ability to integrate CMUTs with electrical circuits, using existing manufacturing methods.

Benchmarking

CMUT performance is benchmarked using pitch-catch and pulse-echo experiments, and operation uniformity is tested in air and in immersion. In a pitch-catch experiment, the transducer is benchmarked using a hydrophone, and in a pulse-echo experiment, the transducer is used both for transmitting and receiving, while comparing the measured signal to the hydrophone response.

Applications

The CMUT-on-CMOS technology and the flip-chip process allows tight integration of CMUTs with front-end electronics, which is necessary for miniature medical imaging devices, such as IVUS.

Electromagnetic Acoustic Transducer

Electromagnetic acoustic transducer (EMAT) is a transducer for non-contact sound generation and reception using electromagnetic mechanisms. EMAT is an ultrasonic nondestructive testing (NDT) method which does not require contact or couplant, because the sound is directly generated within the material adjacent to the transducer. Due to this couplant-free feature, EMAT is particularly useful for automated inspection, and hot, cold, clean, or dry environments. EMAT is an ideal transducer to generate Shear Horizontal (SH) bulk wave mode, Surface Wave, Lamb waves and all sorts of other guided-wave modes in metallic and/or ferromagnetic materials. As an emerging ultrasonic testing (UT) technique, EMAT can be used for thickness measurement, flaw detection, and material property characterization. After decades of research and development, EMAT has found its applications in many industries such as primary metal manufacturing and processing, automotive, railroad, pipeline, boiler and pressure vessel industries.

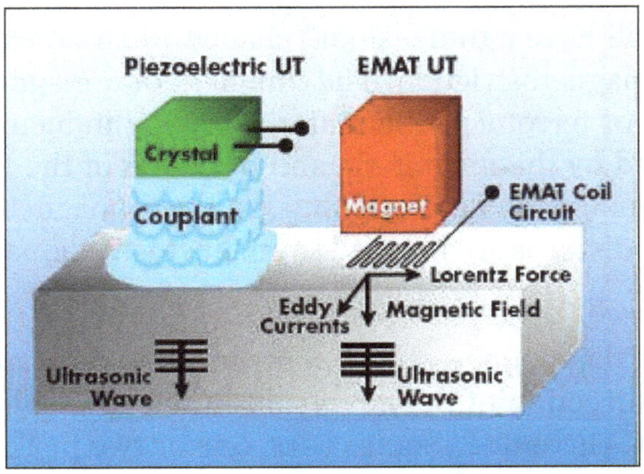

An EMAT ultrasonic transducer (UT) shown with a conventional piezoelectric UT.

Basic Components

There are two basic components in an EMAT transducer. One is a magnet and the other is an electric coil. The magnet can be a permanent magnet or an electromagnet, which produces a static or a quasi-static magnetic field. In EMAT terminology, this field is called bias magnetic field. The electric coil is driven with an alternating current (AC) electric signal at ultrasonic frequency, typically in the range from 20 kHz to 10 MHz. Based on the application needs, the signal can be a continuous wave, a spike pulse, or a tone-burst signal. The electric coil with AC current also generates an AC magnetic field. When the test material is close to the EMAT, ultrasonic waves are generated in the test material through the interaction of the two magnetic fields.

Transduction Mechanism

There are two mechanisms to generate waves through magnetic field interaction. One is Lorentz force when the material is conductive. The other is magnetostriction when the material is ferromagnetic.

Lorentz Force

The AC current in the electric coil generates eddy current on the surface of the material. According to theory of electromagnetic induction, the distribution of the eddy current is only at a very thin layer of the material, called skin depth. This depth reduces with the increase of AC frequency, the material conductivity, and permeability. Typically for 1 MHz AC excitation, the skin depth is only a fraction of a millimeter for primary metals like steel, copper and aluminum. The eddy current in the magnetic field experiences Lorentz force. In a microscopic view, the Lorentz force is applied on the electrons in the eddy current. In a macroscopic view, the Lorentz force is applied on the surface region of the material due to interaction between electrons and atoms. The distribution of Lorentz force is controlled by the design of magnet, and design of the electric coil, the properties of the test material, relative position between the transducer and the test part, and the excitation signal for the transducer.

Magnetostriction

A ferromagnetic material will have a dimensional change when an external magnetic field is applied. This effect is called magnetostriction. The flux field of a magnet expands or collapses depends on the arrangement of ferromagnetic material having inducing voltage in a coil and the amount of change is affected by the magnitude and direction of the field. The AC current in the electric coil induces an AC magnetic field and thus produces magnetostriction at ultrasonic frequency in the material. The disturbances caused by magnetostriction then propagate in the material as an ultrasound wave.

In polycrystalline material, the magnetostriction response is very complicated. It is affected by direction of the bias field, direction of the field from AC electric coil, the strength of bias field, and the amplitude of the AC current. In some cases, one or two peak response may be observed with the increase of bias field. In some cases, the response can be improved significantly with the change of relative direction between bias magnetic field and AC magnetic field. Quantitatively, the magnetostriction may be described in a similar mathematical format as piezoelectric constants. Empirically, a lot of experience is needed to fully understand the magnetostriction phenomenon.

Magnetostriction effect has been used to generate both SH-type and Lamb type waves in steel products. Recently, due to the stronger magnetostriction effect in nickel than steel, magnetostriction sensors using nickel patches are also developed for nondestructive testing of steel products.

Comparison with Piezoelectric Transducers

As an ultrasonic testing (UT) method, EMAT has all the advantages of UT compared to other NDT methods. Just like piezoelectric UT probes, EMAT probes can be used in pulse echo, pitch-catch, and through-transmission configurations. EMAT probes can also be assembled into phased array probes, delivering focusing and beam steering capabilities.

Advantages

Compared to piezoelectric transducers, EMAT probes have the following advantages:

1. No couplant is needed. Based on the transduction mechanism of EMAT, couplant is not required. This makes EMAT ideal for inspections at temperatures below the freezing point and above the evaporation point of liquid couplants. It also makes it convenient for situations where couplant handling would be impractical.
2. EMAT is a non-contact method. Although proximity is preferred, a physical contact between the transducer and the specimen under test is not required.
3. Dry Inspection. Since no couplant is needed, the EMAT inspection can be performed in a dry environment.
4. Less sensitive to surface condition. With contact-based piezoelectric transducers, the test surface has to be machined smoothly to ensure coupling. Using EMAT, the requirements to surface smoothness are less stringent; the only requirement is to remove loose scale and the like.
5. Easier for sensor deployment. Using piezoelectric transducer, the wave propagation angle in the test part is affected by Snell's law. As a result, a small variation in sensor deployment may cause a significant change in the refracted angle.
6. Easier to generate SH-type waves. Using piezoelectric transducers, SH wave is difficult to couple to the test part. EMAT provide a convenient means of generating SH bulk wave and SH guided waves.

Challenges and Disadvantages

The disadvantages of EMAT compared to piezoelectric UT can be summaried as follows:

1. Low transduction efficiency. EMAT transducers typically produce raw signal of lower

power than piezoelectric transducers. As a result, more sophisticated signal processing techniques are needed to isolate signal from noise.

2. Limited to metallic or magnetic products. NDT of plastic and ceramic material is not suitable or at least not convenient using EMAT.
3. Size constraints. Although there are EMAT transducers as small as a penny, commonly used transducers are large in size. Low-profile EMAT problems are still under research and development. Due to the size constraints, EMAT phased array is also difficult to be made from very small elements.
4. Caution must be taken when handling magnets around steel products.

Applications

EMAT has been used in a broad range of applications and has potential to be used in many other applications. A brief and incomplete list is as follows.

1. Thickness measurement for various applications.
2. Flaw detection in steel products.
3. Plate lamination defect inspection.
4. Bonded structure lamination detection.
5. Laser weld inspection for automotive components.
6. Various weld inspection for coil join, tubes and pipes.
7. Pipeline in-service inspection.
8. Railroad and wheel inspection.
9. Austenitic weld inspection for power industry.
10. Material characterization.

Orthomode Transduce

An orthomode transducer (OMT) is a waveguide component. It is commonly referred to as a *polarisation duplexer*. Orthomode transducers serve either to combine or to separate two orthogonally polarized microwave signal paths. One of the paths forms the uplink, which is transmitted over the same waveguide as the received signal path, or downlink path. Such a device may be part of a VSAT antenna feed or a terrestrial microwave radio feed; for example, OMTs are often used with a feed horn to isolate orthogonal polarizations of a signal and to transfer transmit and receive signals to different ports.

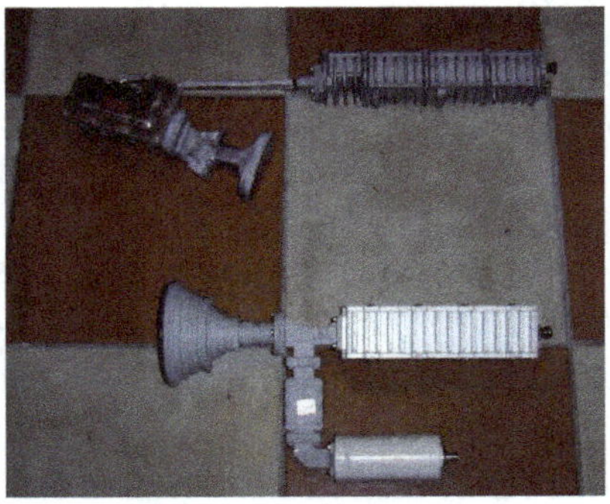

Outdoor unit, includes feed horn, OMT, LNB and BUC.

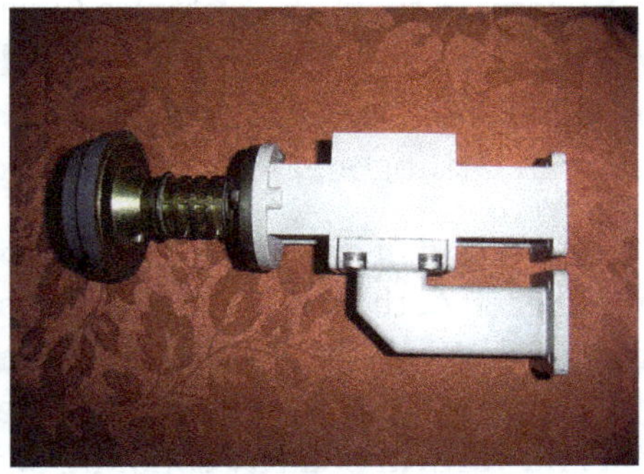

Orthomode transducer.

Orthomode transducer, vertical and horizontal polarity.

Antenna side of OMT.

VSAT and Satellite Earth Station Applications

For VSAT modems the transmission and reception paths are at 90° to each other, or in other words, the signals are orthogonally polarized with respect to each other. This orthogonal shift between the two signal paths provides approximately an isolation of 40 dB in the K_u band and K_a band radio frequency bands.

Hence this device serves in an essential role as the junction element of the outdoor unit (ODU) of a VSAT modem. It protects the receiver front-end element (the low-noise block converter, LNB) from burn-out by the power of the output signal generated by the block up converter (BUC). The BUC is also connected to the feed horn through a wave guide port of the OMT junction device.

Orthomode transducers are used in dual-polarized Very small aperture terminals (VSATs), in sparsely populated areas, radar antennas, radiometers, and communications links. They are usually connected to the antenna's down converter or LNB and to the High Power Amplifier (HPA) attached to a transmitting antenna.

Wherever there are two polarizations of radio signals *(Horizontal and Vertical)*, the transmitted and received radio signal to and from the antenna are said to be "orthogonal". This means that the modulation planes of the two radio signal waves are at 90 degrees angles to each other. The OMT device is used to separate two equal frequency signals, of high and low signal power. Protective separation is essential as the transmitter unit would seriously damage the very sensitive low micro-voltage (μV), front-end receiver amplifier unit at the antenna.

The transmission signal of the up-link, of relatively high power (1, 2,or 5 watts for common VSAT equipment) originating from BUC *(block up converter)*, and the very low power received signal power (μV) coming from the antenna (aerial) to the LNB receiver unit, in this case are at an angle of 90° relative to each other, are both coupled together at the feed-horn focal-point of the Parabolic antenna. The device that unites both up-link and down-link paths, which are at 90° to each other, is known as an Orthogonal Mode Transducer *OMT*.

In the VSAT K_u band of operation case, a typical *OMT* Orthomode Transducer provides a 40 dB isolation between each of the connected radio ports to the feed horn that faces the parabolic dish reflector (40 dB means that only 0.01% of the transmitter's output power is cross-fed into the receiver's wave guide port). The port facing the parabolic reflector of the antenna is a circular polarizing port so that horizontal and vertical polarity coupling of inbound and outbound radio signal is easily achieved.

The 40 dB isolation provides essential protection to the very sensitive receiver amplifier against burn out from the relatively high-power signal of the transmitter unit. Further isolation may be obtained by means of selective radio frequency filtering to achieve an isolation of 100 dB (100 dB means that only a 10^{-10} fraction of the transmitter's output power is cross-fed into the wave guide port of the receiver).

The second image demonstrates two types of outdoor units, a 1-watt Hughes unit and a composite configuration of a 2-watt BUC/OMT/LNB Andrew, Swedish Microwave units.

The following images show a Portenseigne & Hirschmann K_u band configuration, that highlights the horizontal the vertical, and circular polarized wave-guide ports that join to the Feed-horn, the LNB or BUC elements of an outdoor unit.

Terrestrial Microwave

An ortho-mode transducer is also a component commonly found on high capacity terrestrial microwave radio links. In this arrangement, two parabolic reflector dishes operate in a point to point microwave radio path (4 GHz to 85 GHz) with four radios, two mounted on each end. On each dish a T-shaped ortho-mode transducer is mounted at the rear of the feed, separating the signal from the feed into two separate radios, one operating in the horizontal polarity, and the other in the vertical polarity. This arrangement is used to increase the aggregate data throughput between two dishes on a point to point microwave path, or for fault-tolerance redundancy. Certain types of outdoor microwave radios have integrated orthomode transducers and operate in both polarities from a single radio unit, performing cross-polarization interference cancellation (XPIC) within the radio unit itself. Alternatively, the orthomode transducer may be built into the antenna, and allow connection of separate radios, or separate ports of the same radio, to the antenna.

CableFree 2+0 XPIC Microwave Link showing OMT and two ODUs connected to H & V polarity ports.

Characterization

An ortho-mode transducer can be modelled as a 4-port device, 2 of these (H and V) representing the single-polarization ports and the remaining (h, v) embodied by the degenerate modes in the dual-polarized port.

The scattering parameters can be gathered in a 4×4 scattering matrix S, which is symmetrical for a reciprocal OMT (i.e. not including circulators, isolators or active components), thus leaving 10 independent terms for a general lossy device:

$$\mathrm{S} = \begin{bmatrix} S_{HH} & S_{HV} & S_{Hh} & S_{Hv} \\ S_{HV} & S_{VV} & S_{Vh} & S_{Vv} \\ S_{Hh} & S_{Vh} & S_{hh} & S_{hv} \\ S_{Hv} & S_{Vv} & S_{hv} & S_{vv} \end{bmatrix}$$

Of these:

- 4 $(S_{HH} S_{VV,} S_{hh}, S_{vv})$ represent the intrinsic reflection terms of the 4 ports, related to the return loss when all the ports are closed onto ideal loads equal to the port characteristic impedance;
- 2 $(S_{Hh} SV_{v})$ are the main direct transmission terms (from each single-polarization port to the corresponding mode on the dual-polarized port);
- 2 $(S_{HV} S_{Vh})$ represent the cross-polarization discrimination (XPD): from each single-polarization port to the supposedly-isolated mode on the dual-polarized port;
- 2 $((S_{HV} S_{hv})$, model the isolation terms (sometimes referred as inter-port isolation, IPI): between the two single-polarized ports and between the two orthogonal modes at the dual-polarized port.

An ideal OMT exhibits perfect matching (null terms on the diagonal), unitary direct transmission terms and infinite XPD and isolation (null corresponding scattering parameters):

$$\mathrm{S} = \begin{bmatrix} 0 & 0 & 1 & 0 \\ 0 & 0 & 0 & 1 \\ 1 & 0 & 0 & 0 \\ 0 & 1 & 0 & 0 \end{bmatrix}$$

Characterization of a manufactured OMT (considered the device under test, DUT) is usually a delicate matter for both mechanical and theoretical reasons.

Conceptually, if an ideal OMT is available as part of the measurement setup, often named "golden sample", its dual-polarized port can be connected to its counterpart on the DUT, resulting in a 4-port equivalent device with 4 single-polarization ports. The ideal OMT splits the two polarizations at the dual-polarized port into two standard single-polarized ports and such arrangement allows the direct measurement of all the scattering parameters of the DUT (either by using a 4-port vector network analyzer (VNA) or a 2-port one with 2 single-polarized loads used in several combinations).

Such ideal setup is only prone to mechanical uncertainties related to the physical placement and alignment of the dual-polarized ports. A simple misalignment angle $\in$ introduces an artificial path from each polarization to the opposite proportional to $\in$. The phasorial combination of the leakage $S_{Vh}(or\ SHv)$ (or) due to the XPDs of DUT and this artificial loss Sin$\in$ is the actual external measured quantity. If, by proper phase recombination, the two contributions tend to cancel each other, the actual measured XPD can increase to infinity (possible only if $|S_{Vh}|=|\sin\in|$), thus resulting in a huge estimation error.

Depending on the expected XPD of the DUT, mechanical countermeasures should be introduced to guarantee that the artificial measurement uncertainty can be neglected.

Any deviation from this ideal setup, however, introduces errors and uncertainties.

If a dual-polarization matched load is available in place of the ideal OMT, this allows 2×2 measurements from the single-polarization ports, yielding only 2 of the reflection terms (S_{HH} and S_{HH}) and one IPI (S_{HV}). Other measurements aimed at gaining estimations of the other scattering parameters of the DUT involve the dual-polarized port and require additional components, such as dual-polarized to single-polarized transitions or tapers, which are often not matched on at least one of the two polarizations: this creates undesired reflections which propagate through the OMT and combine at the VNA ports thus preventing direct measurements. These issues add to mechanical factors and enhance uncertainties in the measurement procedure.

Due to the increasing demand for high-capacity data links, the exploitation of dual-polarization has fostered research in design and characterization of OMTs to overcome the practical difficulties. The literature concerning OMT modelling and practical characterization consists of works both by academic organizations such as the National Research Council (Italy), Marche Polytechnic University and European Space Agency and likewise by industrial teams such as CommScope and Siae Microelettronica with immediate impact on products for modern dual-polarized telecommunication systems, for instance in terrestrial microwave backhauling.

References

- Transducer-electronics, technology: britannica.com, Retrieved 6 August, 2019
- "General Description and Advantages of cmuts". Stanford University. Archived from the original on 20 July 2011. Retrieved 7 February 2011
- Types-of-transducer: circuitglobe.com, Retrieved 7 January, 2019
- Liu, Jessica; Oakley, Clyde; Shandas, Robin (2009). "Capacitive micromachined ultrasonic transducers using commercial multi-user mumps process: Capability and limitations". Ultrasonics. 49 (8): 765–773. Doi:10.1016/j.ultras.2009.06.003. ISSN 0041-624X. PMC 2783530. PMID 19640557
- Resistive-transducer: circuitglobe.com, Retrieved 8 February, 2019
- Syme, Jim (26 August 2014). "Back to Basics in Microwave Systems: Cross-Polar Discrimination". Retrieved 6 December 2016
- Capacitive-transducer: circuitglobe.com, Retrieved 9 March, 2019
- Piezoelectric-transducer: instrumentationtoday.com, Retrieved 10 April, 2019
- Accelerometer: instrumentationtoday.com, Retrieved 11 May, 2019
- Pressure-transducer: instrumentationtoday.com, Retrieved 12 June, 2019

Types of Sensors 3

- Humidity Sensors
- Biosensors
- Optical Sensors
- Position Sensors
- Light Sensors
- Hall Effect Sensor
- Oxygen Sensor
- Wheel Speed Sensor
- Torque Sensor
- Throttle Position Sensor
- Soil Moisture Sensor
- Active Pixel Sensor
- Tachometer
- Air Pollution Sensor

Some of the different types of sensors are humidity sensors, biosensors, optical sensors, position sensors, light sensors, Hall effect sensors, oxygen sensors, wheel speed sensors, torque sensors, throttle position sensors, soil moisture sensors, active pixel sensors, etc. This chapter discusses in detail all these types of sensors.

Humidity Sensors

Humidity is one of the most commonly measured physical quantities and is of great importance in a wide variety of commercial and industrial applications, including those associated with building ventilation control, clean rooms in the semiconductor and automotive industries, environmental chambers for the testing of electronics, industrial drying, and process monitoring in the chemicals, electronics, food/beverage, pharmaceutical, cosmetics and biomedical analysis industries.

Humidity is defined as a measure of the water vapor present in a gas. Two common parameters in associated with humidity measurement are absolute humidity and relative humidity. Absolute humidity is referred to a measure of the mass of water vapor in a unit volume. Similarly, relative humidity is defined as the ratio of the water-vapor pressure present to the water-vapor pressure required for saturation at a given temperature. The relative humidity is related to ambient temperature. Water vapor is a natural component of air, and the relative humidity (RH) of the water vapor and air mixture is defined as the ratio of the mass of water vapor in a unit volume compared to the mass of water vapor which that volume could hold if the vapor were saturated at the mixture temperature. RH is typically expressed as:

$$RH = \frac{p_w}{P_s} \times 100\%$$

where, P_w and P_s are the vapor and saturation pressures, respectively.

Another commonly measured parameter associated with relative humidity is the dew point, T_d (°C) which defines the temperature below which condensation of water occurs. The expressions and definitions of the wet-bulb and dew-point temperatures are plotted in figure. Having measured the values of P_w and P_s, it is clear that the corresponding relative humidity value can be determined from equation above.

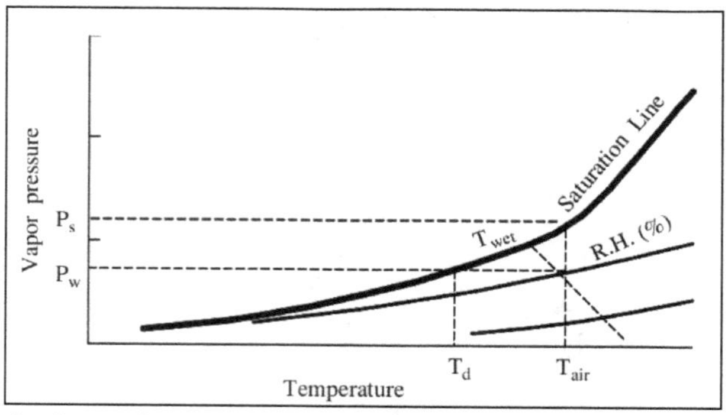

Psychrometric chart, where T_{air} is the ambient air temperature. If the temperature drops below the dew point, T_d (°C), condensation of water occurs.

Humidity measurement is generally conducted using a psychrometer. Figure presents a schematic illustration of a psychrometer (IMAG-DLO, IMax) comprising two thermometers, which measure the dry-bulb and wet-bulb temperatures, respectively. The vapor pressure, P_w, can then be calculated from:

$$P_w = P_s - P_t A\left(T_{wet} - T_{air}\right)$$

where P_s is the saturation vapor pressure at T_{air} (dry-bulb temperature), P_t is the total medium pressure T_{wet} is the measured wet-bulb temperature, and A is the psychrometer constant, which is a function of the psychrometer geometry and the air flow rate.

It can be shown that the approximated thermodynamic psychrometer constant (A) has a value of 5.68 at 0 °C, 6.44~6.46 at 20 °C, 6.28~6.45 at 40 °C, and 5.63~6.43 at 60 °C. This constant is temperature-dependent and varies more greatly with relative humidity under high temperature conditions. The thermodynamic properties are calculated using the perfect-gas relationships of moist air. Although psychrometers have the disadvantage of requiring regular maintenance, their operational principle is simple and unique. Hence, their use is invaluable when calibrating other air humidity sensors.

Optical Humidity Sensors

Fiber-optic humidity sensors are based on the colorimetric interaction of materials immobilized on the surface of the fiber core or its cladding in the humidity-sensing section. The sensing mechanism relies on a humidity-induced refractive index change in the materials, which causes the transmitted optical intensity through the sensing section to vary as a function of the relative humidity.

Kharaz and Jones presented an optical-fiber humidity sensing system based on the colorimetric interaction of cobalt chloride with water molecules. As shown schematically in figure, a 50 mm length of the fiber cladding was removed and replaced by a cobalt chloride-gelatin thin film. While maintaining a constant temperature of 36 °C, it was shown that variations of relative humidity were reflected as changes in the spectral absorption of the sensor in the wavelength range of 600–740 nm. Meanwhile, a second wavelength beyond this range served as an intensity reference. The relative intensity of the signals was obtained by dividing the intensity of the light at the 670 nm wavelength by that at a wavelength of 850 nm. The average humidity sensitivity was found to be 0.001~0.007 a.u./%RH.

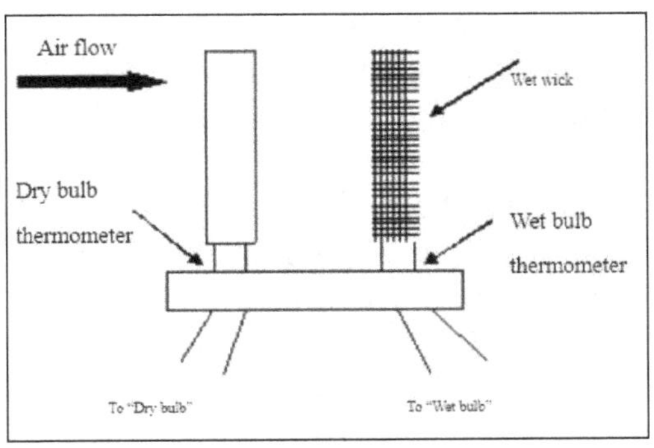

Schematic illustration of psychrometer. The wet bulb thermometer is kept moist by a cotton wick placed in a water reservoir, and is used to measures T_{wet}. The dry bulb thermometer measures the ambient air temperature T_{air}.

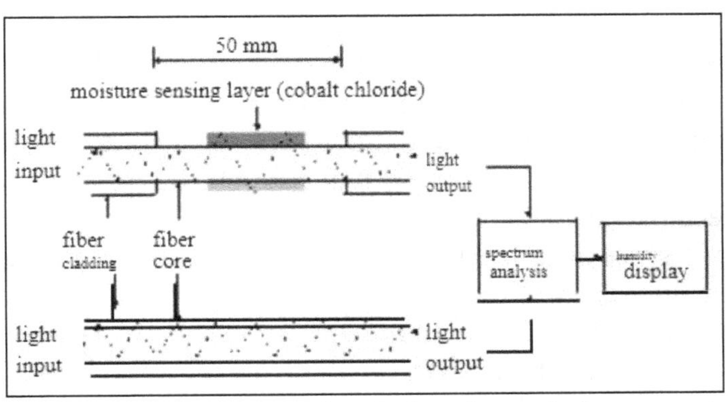

Construction and operation of an optical.

A 50 mm length of the fiber cladding is removed and replaced by a cobalt chloride-gelatin thin film. When the temperature is maintained at 36 °C, a change of relative humidity affects the spectral absorption of the sensor in the wavelength region 600–740 nm. Meanwhile, a second wavelength outside this region provides an intensity reference. The relative intensity of the signals is obtained by dividing the intensity of the light at 670 nm by that of the light at 850 nm. The average humidity sensitivity is 0.001~0.007 a.u./%RH at different operating regions.

In an alternative approach, a hydrophilic gel (agarose) was deposited on the thinner zone of a biconically tapered single-mode optical fiber. Using this arrangement, 6.5 dB of attenuation was obtained as the relative humidity was changed from 30% to 80%. However, an apparent hysteresis of 0.5 dB was also observed.

Optical techniques also make use of the absorption wavelength of water. When light is passed through a gas, the absorption of certain wavelengths gives rise to a significant change in both the polarization and the amplitude of the incident optical signal. For example, Schirmer et al. presented a fiber-coupled diode laser absorption spectrometer modulated with 750 ± 001 GHz for the detection of humidity at different frequencies. Once the background noises had been subtracted, it was determined that the absorption signals were proportional to the humidity volume fractions with a sensitivity of 0.01105 V/ppm.

Films of PVA (polyvinyl alcohol)/H_3PO_4 containing dyes (CV {crystal violet}/MB {methylene blue}) were applied to optical humidity sensing by Somani et al. An apparent spectrum shift was observed when the relative humidity was increased from 9% RH to 100% RH.

Ha et al. presented a mechanical-optoelectronic humidity sensor comprising an LED (light-emitting device), a photodiode, and a mechanical system.8 In this device, one end of the hair bunch is fixed, while the other end is connected to a thin metal sheet, which contains a window and is connected to a spiral spring. A humidity change causes a contraction or expansion of the hair, which then pulls the metal sheet up or down. The motion of the sheet relative to the fixed metal target changes the window area, which in turn causes a change in the LED light intensity detected at the photodiode and a corresponding change in the output photocurrent. Compared to capacitive or resistive humidity sensors, the mechanical-optoelectronic humidity sensor demonstrates an improved long-term stability and a reduced temperature dependency.

Researchers reported a fiber-optic humidity sensor based on the moisture-dependent absorption of light by using a phenol red doped polymethylmethacrylate (PMMA) film over a small portion of the core of the plastic clad silica (PCS) fiber with a lower refractive index. An ingenious U-shaped probe was designed to decrease the angle of incidence of the ray at the core-film interface in the bent region and hence to increase the transmission of light from the core to the sensor. The response time of this sensor is about 5 sec while operating from 20 to 80% RH, which demonstrated a rapid response for optical humidity sensors. Shukla et al. reported nano-structured magnesium-oxide films by using sol–gel techniques to sense the moisture surrounding a similar U-shaped probe. In their study, optical fibers were utilized to sense adsorption of humidity on nanosize MgO films coated on U-shaped glass rods while a light guide received the input optical power at one end and delivered the output optical power at the other end. The change in the output optical power was shown to be highly related with the humidity. Besides, polymer-based optode membranes doped with fluorescence were also used for sensing of relative humidity, and demonstrated a high detection resolution up to 0.56%.

Sorli et al. developed optical humidity sensors based on the cooled mirror principle using optical detection techniques. The sensors were composed of a reflective sensor and a Peltier module, which controlled the temperature to reach the dew point and to evaporate the condensate water. A LED and a photo-detector were used to measure the water droplets. The water contents ranging between 80 and 95% were measured with high accuracy. Drew et al. demonstrated a humidity sensor using vapor-luminescent platinum (II) double salt materials. Luminescence spectra for platinum (II) double salt materials showed that the intensity and the wavelength maximum corresponded to the concentration of water vapor surrounding the materials. The excitation light was generated by mounting a blue LED at the end of one of the bifurcated fibers and the collection end of the bifurcated fiber was connected to a spectrophotometer for the analysis of the spectrum of the luminescent platinum (II) double salt materials. Thus water vapor could be detected successfully using this approach.

Photo-acoustic (PA) humidity sensing systems using laser diodes were also developed. Due to the dependence of the photo-acoustic signals on various physical and chemical properties of the measured gas samples, minimum detectable concentration of water vapor up to sub-ppm level could be achieved using the developed diode-laser-based PA water vapor detection systems (DPWDS).

Although optical humidity sensors have a high sensitivity, several drawbacks hinder their practical application, including hysteresis, typically varying from 0.5% to 1% RH, their bulky nature, the requirement for frequent mirror decontamination, their instability under continuous use, a relatively high power consumption, the requirement for some form of temperature compensation mechanism if precise humidity measurements are to be obtained, and finally, their relatively high costs. Notwithstanding these limitations, the capability of optical humidity sensors to track low-level moisture levels is superior to that of their counterparts and hence, these devices are generally employed in laboratory environments.

Gravimetric Humidity Sensors

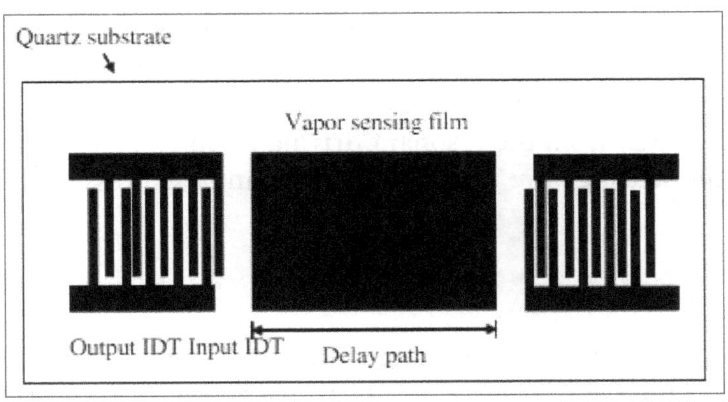

Schematic illustration of a SAW sensor. The sensor creates a series of SAWs which travel across the sensing film to the output IDT where they are converted back into an electrical signal.

Surface acoustic wave (SAW) devices have been widely used in humidity measurement applications. Wohltjen presented a SAW sensor with the ability to measure vapor (humidity) with a greater sensitivity than bulk wave sensors. In a SAW sensor, surface acoustic waves are launched along the delay path of the SAW sensor via the reverse piezoelectric effect when an RF signal is applied at the center frequency to the input interdigital transducer (IDT). This creates a series of SAWs, which travel across the vapor sensing film to the output IDT, where they are then converted back to an electrical signal. Hence, the relative change (Δυ) in the SAW phase velocity (υ_o) caused by the mass density changes induced by vapor absorption can be given approximately as:

$$\frac{\Delta v}{v_o} = -\frac{\pi h v_o^2}{2\lambda} \Delta\rho(A_x^2 + A_y^2 + A_z^2)$$

where h is the thickness of the film, Δ is the change of the v mass density, λ is the acoustic wavelength, and Ai ($i = x, y, v or z$) are the normalized displacement amplitudes of the SAW.

Hence, the relative humidity can be determined by measuring the relative change of the phase velocity. The relative changes in the phase velocity and frequency are related by the expression given in equation below:

$$\frac{\Delta v}{v_o} = \frac{\Delta f}{f_o}$$

where Δf is the frequency shift caused by the vapor absorption and f_o is the SAW frequency.

A typical SAW humidity sensor is operated using RF amplifiers with a coupler. The oscillation frequency is monitored using a frequency counter. Although these sensors are relatively sensitive, oscillator instability can pose problems in practical applications. Typically, the ratio $\Delta f/f_o$ of the SAW humidity sensor increases from 0 to 22 ppm as the relative humidity changes from 0% RH to 85% RH when a polyethynylfluorenol (PEFL) sensing membrane is used. It is clear that this sensitive membrane yields a high humidity sensitivity. It is noted that cellulose acetate films, hexam-ethyldisiloxane (HMDSO) and polymer-like hydrogenated amorphous carbon (a-C H) films, Al_2O_3 ceramic thin film, polyvinyl alcohol (PVA) film and other polymer films have also been reported as sensitive membranes.

Radeva et al. presented an alternative type of gravimetric humidity sensor. In his study, the sensitive detection capability ($10^{-8} \sim 10^{-11} g$) of a quartz crystal microbalance (QCM) was exploited to perform the characterization of vapors. The operation of this sensor is based on the shift of the resonant frequency (Δf) away from the resonant driving frequency (f_r) when an additional mass (Δ*m*) is deposited on the sensing film. This dependency can be expressed as:

$$\frac{\Delta f}{f_r} = -\frac{\Delta m}{M}$$

where M is the mass of the quartz piezoelectric substrate. Thin (0.12 m) fullerene layers are deposited on either side of the QCM to serve as adsorptive layers for water molecules. The mass of water molecules absorbed by the fullerene layer causes a decrease of the resonant frequency. The number of water molecules absorbed depends on their concentration in the environment (i.e. the relative humidity). These sensors also demonstrate a quick response (1.0 sec) to cyclic changes in the air humidity.

Glück et al. proposed a cantilever-type resonator utilizing a stratified structure with a piezoelectric polymer layer (polyvinyl-difluorene, PVDF). A cantilever vibration is induced by the expansion or contraction which takes place when an electrical signal is applied to the electrodes placed on either side of the PVDF. The adsorption of water vapor results in a change of the mass, which causes a shift in the resonant frequency of the beam. In the range 0–90% RH, a frequency shift of 30 Hz was observed at a resonant frequency of 1.39 kHz.

In general, gravimetric humidity sensors provide a high sensitivity, a rapid response time and a simple experimental setup. However, they generally require expensive driving and detection electronics.

Capacitive Humidity Sensors

Approximately 75% of the humidity sensors on the market today are based on the capacitive technique. Capacitive humidity sensors offer several advantages, including very low power consumption and high output signals. Using these sensors, the ambient relative humidity level is measured by detecting moisture-induced changes in the dielectric constant of a hygroscopic layer. Typically, capacitive humidity sensors are composed of two interdigitated electrodes (IDE) covered by a dielectric layer, which is sensitive to humidity changes. The measured capacitance values demonstrate a nonlinear relationship with the relative humidity. Story et al. employed a polyimide film as the moisture sensitive layer. Although a thin polyimide film (2 m) yields a higher sensitivity (27 fF/%RH), serious hysteresis occurs at high values of RH. Hysteresis is a common problem for virtually all kinds of capacitive humidity sensors due to the slower diffusion time of moisture sensitive films while dehumidifying.

Other materials such as PMMA, porous ceramics, porous silicon, porous silicon carbide and hygroscopic polymers have also been utilized as humidity sensitive materials. Nahar et al. presented highly sensitive humidity sensors using porous Al_2O_3 thin films. Laville and Pellet proposed interdigitated humidity sensors incorporating a plasmaetched polyimide sensing layer. Thin polyimide films exhibit a low humidity sensitivity (4 pF/%RH) at 0% < RH < 70% and a high sensitivity (75 pF/%RH) at 70% < RH < 90%. It has been reported that thin sensing layer films increase the sensitivity of humidity sensors. To reduce the serious hysteresis problems, Fürjes et al. integrated an internal heater over the moisture sensing layer. However, the hysteresis phenomenon still remains problematic. Porous silicon as a miniaturized humidity sensor material was reported. For example, Björkqvist et al. developed a capacitive humidity sensor using thermally carbonized porous silicon (TC-PS) at room temperature. In their study, long-term stability of porous silicon was improved by thermal oxidation since a stable surface of thermally carbonized PS was formed. In spite of good sensitivity over a wide range of relative humidity and excellent repeatability, the hysteresis above 60% of relative humidity was still observed.

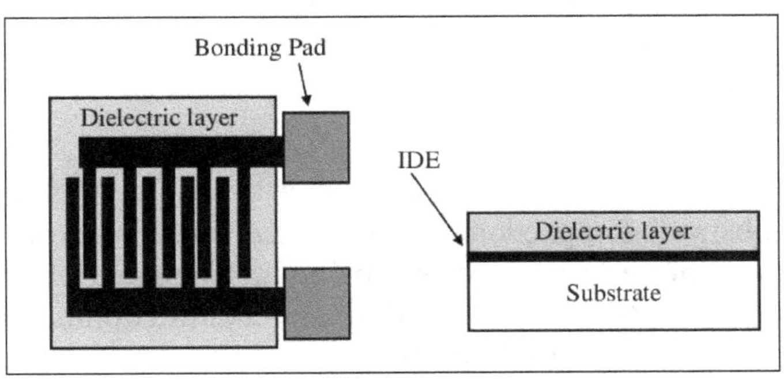

Capacitive humidity sensors with interdigitated electrodes. Typically, capacitive humidity sensors are composed of two interdigitated electrodes (IDE) covered by a dielectric layer, which is sensitive to humidity change.

For instance, Connolly et al. used SiC membrane and Al electrodes rather than Au which was commonly used in conventional humidity sensors such that the fabrication was more cleanroom friendly and compatible with CMOS process. Polymer materials were also used for humidity sensing. For example, Matsuguchi et al. developed a capacitive humidity sensor based on a poly(methyl methacrylate) (PMMA) cross-linked with divinylbenzene (DVB). Thanks to the rigid cross-linked structures in the polymer, the irreversible increase in volume of the sensing layers caused by swelling was prevented, which improved the long-term stability of the capacitive humidity sensors.

Typically, ceramics have the advantages of high mechanical strength and the ability to function correctly at high temperatures. Ceramics used in humidity sensing applications include $MgCr_2O_4$, TiO_2 and Al_2O_3. Compared to ceramics, polymers are more readily applied to sensing devices and are more compatible with standard IC fabrication technology. Typical hygroscopic polymers include cellulose acetate, polymethyl methacrylate, polyimide, plasmapolymerized hexamethyldisilazane (HMDSN) and polyethersulphone. For example, Gu et al. reported a capacitive humidity sensor integrated with a polysilicon heater using a CMOS process. The humidity sensing material, polyimide, was deposited by a post-processing step after the standard CMOS fabrication. The integrated heater was shown to be able to improve the speed of moisture desorption and eliminate the volatile impurity. A quick response time (10 s) was observed in their study.

Systematic investigations have shown that the ambient humidity has a significant influence on both the electrical and mechanical properties of polyimide. When using a bulk polyimide film as the sensing material, the humidity is generally determined from capacitance measurements and a good linear relationship is obtained between the output signal and the relative humidity. A change of 25–28% normalized capacitance has been reported as the RH value is increased from 0–100%. Some modified polyimide films using 3-dimensional micro-structures to enlarge the active surface and to improve the measuring sensitivity have also been reported.

Other designs of capacitive humidity sensors were developed in recent years. For example, Chakraborty et al. proposed a moisture sensitive field effect transistor (FET) using $SiO_2/Si_3N_4/Al_2O_3$ gate structures. Harpster et al. developed a passive wireless humidity sensor consisting of an electroplated copper coil for wireless operation and a polyimide film for humidity sensing. A divinyl siloxane benzocyclobutene (BCB) thin film was used as a moisture sensing layer with a sensitivity of 0.1 pF/%RH and a fast response time of 650 ms. Humidity sensors based on nano-structured carbon films produced by supersonic cluster beam deposition were also presented. A sensitivity of 0.1–0.5 pF/%RH in the relative humidity range 10–70% was reported. The nano-structured carbon films were obtained using pure helium gas as the buffer gas in the pulsed microplasma cluster source (PMCS).

Recently, emerging MEMS (Micro-electro-mechanical-systems) technology and micromachining techniques have had a substantial impact on the miniaturization of sensors. These techniques have facilitated the development of new micro-scale sensing instrumentation capable of accessing information in a rapid and accurate fashion. Significantly, the functionality and reliability of these micro sensors can be increased through their integration with mature logic IC (integrated circuit) devices and with other sensors. Many previous studies have reported the development and application of MEMS technologies to the fabrication of micro-scale humidity sensors. For example, Lee et al. developed a novel fabrication process for Pt resistor temperature detectors (RTD) and micro-cantilevers covered with a water-absorbent polyimide layer. These devices were then incorporated in an integrated humidity sensor as a micro temperature sensor and a micro capacitive humidity sensor, respectively. Integrated micro humidity sensors provide the capacitance measurement required for the instantaneous sensing of humidity.

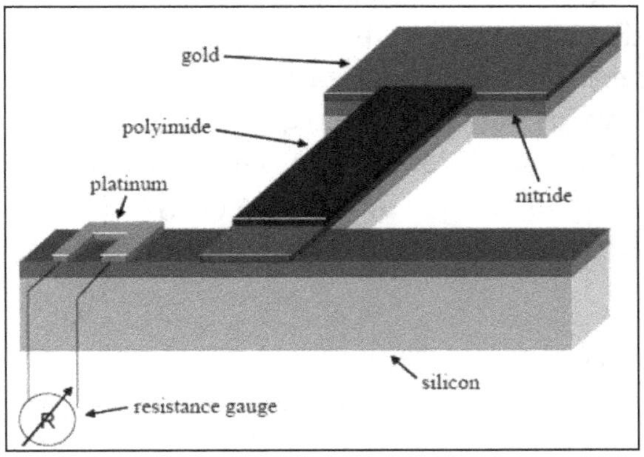

Schematic representation of the integrated humidity sensor and temperature sensor. Absorption of water molecules causes the upper layer of the cantilever to expand, hence inducing a surface tensile stress. As a result, the cantilever bends up, giving rise to a measurable change in the capacitance.

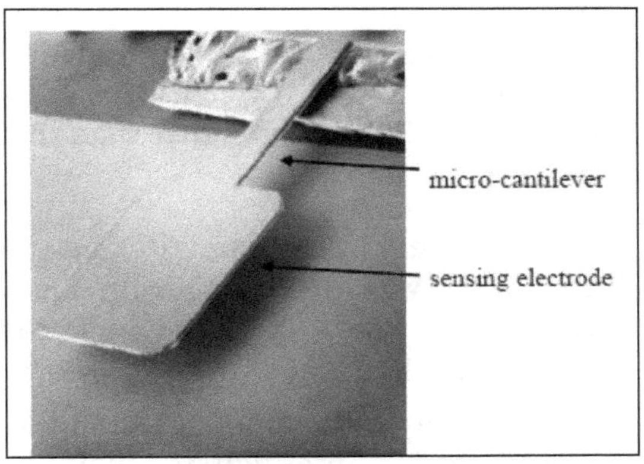

SEM photographic image of the cantilever fabricated for the integrated humidity sensor.

Meanwhile the integrated micro temperature sensors permit temperature compensation of the humidity measurement by providing temperature signals to the humidity transducers. Having compensated for temperature drift, the humidity value can be precisely determined from the measured resistance and capacitance signals. The experimental data indicated a high stability ($< \pm$ 0 8%), a low hysteresis (1.9% RH) at high humidity, a high sensitivity (2.0 nF/%RH), and a rapid response time (1.10 sec).

In the capacitance-type integrated humidity sensor, a movable electrode is fabricated by coating a microfabricated cantilever with a material which absorbs water when exposed to humid conditions. Meanwhile, the second electrode of the capacitor is fixed on a glass substrate. A gold layer is electron-beam evaporated directly on both the moveable cantilever part of the sensor and on the stationary glass substrate. The absorption of water molecules causes the upper layer of the cantilever to expand, and hence induces a surface tensile stress. This causes an upward bending of the suspended structure away from the glass substrate, which changes the capacitance between the two structures.

Previous studies have noted that capacitance-type sensors with a polymer sensing layer are liable to exhibit humidity hysteresis problems. Ideally, a sensor should follow the same capacitance path as the humidity is either increased or decreased. However, in practice, a small degree of hysteresis is exhibited for most humidity sensors, i.e. the capacitance paths for increasing and decreasing humidity differ to some extent. In Lee's investigation, the temperature was maintained at a constant 25 °C while the humidity was increased from 45% RH to 95% RH over a period of 30 minutes and then dehumidified to 45% RH at the same rate. The results revealed that the humidity sensor has a hysteresis of 1.9% RH at high relative humidities (>65% RH) and a value of 0.9% RH at low relative humidities (<65% RH). It is noted that these values are smaller than those reported in previous studies.

In general, capacitive humidity sensors are the most commonly adopted commercially-available humidity sensors. Despite their inherent disadvantages such as hysteresis, slow response time, low output signals (pF) and temperature-dependency, they continue to be the subject of intensive development efforts due to their high sensitivity and low fabrication costs.

Resistive Humidity Sensors

The transduction mechanism of resistive humidity sensors involves the changes in conductivity caused by the adsorption of water vapor. Typically, sensors of this type utilize three major categories of materials: (1) ceramics, (2) polymers and (3) electrolytes. The basic configuration of these sensors resembles that of capacitive humidity sensors other than that the dielectric layers are replaced by conductivity-sensitive layers.

Many previous studies utilized metal oxides with a hubnerite and spinel structure, such as: WO_3, $PbWO_4$, MoO_3, $ZrTiO_3$, $LaFeO_3$, $ZnWoO_4$, WO_3-Y_2O_3 composite, ZnO-Y_2O_3 composite, K_2O doped $Ba_{0.5}Sr_{0.5}TiO_3$, $Sm_{1-x}Sr_xCrO_3$ ($0<x<1$), $La_{1-x}Sr_xFeO_3$ ($0.1<x<0.4$), $MMoO_4$ (M=Ni, Cu, Pb), $Mg_{0.5}Cu_{0.5}Fe_{2-x}M_xO_4$ (x=0 or 0.2, M=La or Ga), Li-Fe_2O_3, $MnWO_4$, $NiWO_4$, $CoWO_4$, $ZnCr_2O_4$, $MgCrO_4$, TiO_2, $MgAl_2O_4$, hematite, ZnO, Al_2O_3, etc. For example, Qu and Meyer presented a thick-film porous $MnWO_4$ ceramic layer sandwiched by interdigitated metal films. Sundaram and Nagaraja used $PbWO_4$ and WO_3 composites for humidity sensing. High sensitivity and quick response were observed in their study. Composites of $MMoO_4$ (M = Ni, Cu, Pb) and $MMoO_3$ were also reported for their good humidity sensing characteristics. A ceramic thick-film humidity sensor with good humidity sensitivity was demonstrated using an emulsion of titania powders by a low-speed spin-coating technique. Although possessing a humidity dependency, the resistance of these sensors generally decreases with ambient temperature. Temperature effects can generate a signal drift, which represents a serious problem for this type of humidity sensor.

Ying et al. presented a humidity sensor based on $TiO_2 - K_2O$–$LiZnVO_4$ ceramic thin films prepared using the sol–gel technique. Excellent humidity sensitive properties were attained over the range of 10–90% RH. Sol–gel $BaTiO_3$ sol–gel silica films synthesized by block copolymers, sol–gel strontium-doped lead-zirconium-titanate ($PS_xZ_{0\,3}T_{0\,7}O_3$, x= 0.1–0.5) and sol–gel chromium titanate oxide ($Cr_{2-x}Ti_xO_3$, x= 0.05–0.4) were also used as moisture sensing layers. It was found that the control of dimension, shape and distribution of pores in these materials was crucial for the fabrication of the mesostructured thin-film humidity sensors. Using di-block, tri-block or star-block, 2-D hexagonal mesoporous phases in silica were formed and good performances in humidity sensing were exhibited. Neri et al. reported the low temperature sol–gel synthesis of chromium titanate oxide (CTO) for the sensing material of resistive humidity sensors. The influence of the doping level of the base material on the humidity sensing performance was also explored. Using sol–gel technique from alkaoxides, Ansari et al. prepared $PS_xZ_{0.3}T_{0.7}O_3$ (x= 0.1–0.5) nano-particles calcined at different temperatures from 500 to 700 °C. While increasing Sr doping level (0.1 M–0.5 M), the band gap in current-voltage (I-V) characteristics was reduced. 0.5 M Sr doping concentration was found to be the optimum condition suitable for relative humidity sensing in the range of 20–100% RH which gave rise to high sensitivity and low barrier height. However, the problem of temperature drift was still evident. Arshak et al. developed a thick-film resistive humidity sensor using a combination of MnO, ZnO and Fe_2O_3. It was shown that this sensor yielded a high linearity, lower hysteresis (0.35% RH) and a fast response time (11–28 sec). Varghese et al. reported the effect of pore size and uniformity on sensing performance for nanoporous Al_2O_3 films. It was reported that the response of the material to humidity was a strong function of pore size and operating frequency. A well-behaved change in impedance more than three orders of magnitude variation over 20% to 90% RH was reported for nanoporous Al_2O_3 films with an average pore size of 13.6 nm.

Due to their ease of fabrication, low cost, and adjustable properties, polymer materials have significant potential for application to humidity sensors. Barkauskas developed a resistive humidity sensor using a sensing film prepared from PVA (polyvinyl alcohol) and graphitized car-bon black disperse phase. The experimental data revealed a sensitivity of 8 ohm/%RH at room temperature and 5 ohm/%RH at 100 °C. However, the response time of this sensor is 45 sec, which is slower than that of the ceramic-type sensor. Wang et al. demonstrated a humidity sensor using Nafion in different ionic forms (H^+, Li^+ and Na^+) prepared using casting techniques. It was shown that the content of the cation in the Nafion has a significant influence on the sensing properties of the film. Su et al. presented a resistive humidity sensor fabricated by thick film techniques using poly (2-acrylamido-2-methylpropane sulfonate) (poly-AMPS) modified with tetraethyl orthosilicate (TEOS) as the sensing material. This sensor was simple to fabricate and exhibited a reduced degree of hysteresis (<2%), good linearity (R2=0 9989) at humidity levels in the range 30–90% RH, long-term stability and satisfactory resistance to high humidity atmospheres (e.g. 95% RH). Fratoddi et al. investigated the resistive-type humidity sensors based two poly(monosubstituted) acetylenes, namely poly(N,N-dimethylpropargylamine) (Pd-PDMPA) and poly(propargylalcohol) (PPOH). These sensors exhibited detectable responses to relative humidity as low as 2% and a variation of five orders of magnitude in the relative humidity range 0–90%.

Ceramic/polymer composite thin films have also been used to optimize the performance of resistive humidity sensors. For example, Feng et al. prepared SiO_2/Nafion composite thin films using casting and dipcoating methods. The performance of Nafion film sensors is determined by the migration of hydrogen ions through the film. At high RH, the total impedance includes the contribution of a faradic impedance at the electrode-film interface. The use of a SiO_2/Nafion composite film permits a significant reduction in the faradic impedance, hence yielding an improved linearity of the sensor response. Chou et al. investigated a humidity sensor with porous ceramics fabricated from ceramic fiber, kaolin and sodium salt of carboxylmethyl cellulose (CMC).

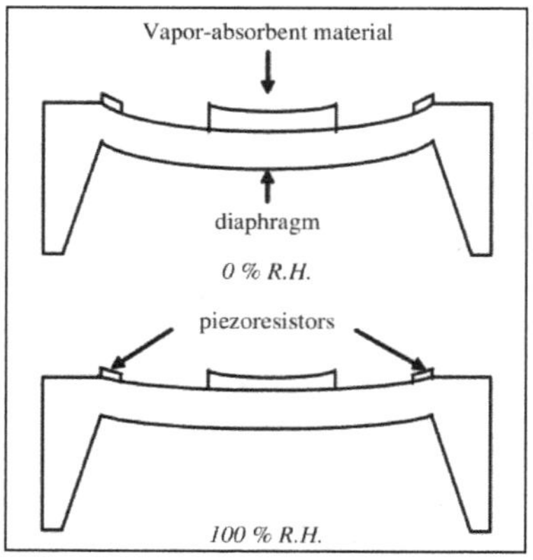

General layouts of piezoresistive humidity sensors. The membrane-type device uses piezoresistors placed on top of the diaphragm. As the humidity increases, the resulting humidity-dependent volume change of the polyimide layer prompts a deformation of the polyimide-bridge substrate bimorph and therefore leads to a bending of the plate, which is then transformed into an output voltage by an integrated piezoresistive bridge.

The conductivity was shown to change by 4–5 orders of magnitude when the RH varied from 10 to 90%. Furthermore, the response time was reported to be 5–8 min for RH values of 10 to 90%. Wang et al. reported a humidity sensing composite material composed of nano-crystalline $BaTiO_3$ and polymer polystyrene sulfonic sodium (PSS). $BaTiO_3$/PSS humidity sensors possess good sensitivity and linearity characteristics at RH ranges of 33–98%, and exhibit a maximum humidity hysteresis of 8% RH. Additionally, the response and recovery times of these sensors are 50 and 120 sec, respectively. This performance is clearly a significant improvement over that of previous polymer sensing materials. Su and Tsai used a composite material of nano-sized SiO_2 and poly(2-Acrylamido-2-methylpropane sulfonate) (poly(AMPS)) to fabricate a resistive humidity sensor. The sensor showed a negligible hysteresis and fast response upon humidification and dehumidification. The applications of nano-sized materials significantly improved the hysteresis.

Sakai et al. developed resistive-type humidity sensors using humidity-sensitive and water-resistive polymetric materials such as quaternized vinylpyridine and styrene copolymers, partially quaternized polyvinylpyridine, polytetrafluoroethylene grafted with quaternized polyvinylpyridine, polyvinylpyridine crosslinked with dibromobutane, poly(2-hydroxy-3-methacryloxypropyl-tri-methyl ammonium chloride) cross-linked with hexam-ethylene diisocyanate, simultaneously crosslinked and quaternized polychloromethyl styrene (PCMS). These sensors demonstrated excellent water resistivity and high sensitivity to humidity. Gong et al. proposed humidity sensors using cross-linked polyelectrolyte prepared from mutually reactive copolymers containing phosphonium salt. These sensors showed average resistances of 725, 43 and 3.1 k at 30, 60 and 90% RH respectively. They were also found to be comparatively resistant to water. Lee et al. proposed two types of resistive humidity sensors using epoxy resin containing quaternary ammonium salts and polyelectrolytes based on new-type mutually cross-linkable copolymers. The impedance of the latter type varies from 755 to 2.52 k in the humidity range of 30% to 90% RH. Cross-linking techniques provide an efficient means of improving the water durability and of providing a long-term stability (300 days). Furthermore, these devices have a rapid response time, typically less than 75 sec when the humidity is changed abruptly from 33 to 94% RH. Using in-situ synthesized inorganic/organic nanocomposites of sodium polystyrenesulfonate (NaPSS) and ZnO, Li et al. developed thin film humidity sensors with four-orders of magnitude sensitivity over 11–97% RH, small hysteresis (less than 2% RH) and quick response time (absorption: 2 s, desorption: 2 s). A conductimetric humidity sensor based on proton conducting perovskite oxide ($Ba_3Ca_{1.18}Nb_{1.82}O_{9-\delta}$, BCN18) was also reported. Dense and porous samples of an oxygen deficient perovskite oxide were fabricated by sintering powder compacts of calcined powders. It was found that porous material had 5–6 times faster time response than that for the dense material. The relatively rapid response of the porous material suggests a promising conductimetric humidity sensor based on porous BCN18 as a good sensing material. Porous silicon (PS)-based metal-insulator-semiconductor (MIS)-structures have been used as moisture sensing materials. The mechanism of humidity sensing has been studied. The current-voltage (I-V) characteristics of these materials could be used for humidity sensing.

Typically, resistive humidity sensors offer the advantages of simple electrical circuits, high sensitivity and good linearity. However, humidity hysteresis is a problem when these sensors are employed on a periodic rather than continuous basis. Furthermore, the sensing materials have a relatively slow response and the long-term stability of the device can only be guaranteed by a continuous operation and by attentive maintenance. Although humidity sensitive materials are still under development, commercially-available resistive humidity sensors are generally made of polymer or porous ceramics. Thick film technology is desirable for the mass production of resistive humidity sensors on account of its cost efficiency, robustness and flexibility in device design.

Piezoresistive Humidity Sensors

Piezoresistive materials have found widespread application in human-made sensors. Nowadays, transduction from air humidity via mechanical domain is a mature sensing technique. Typically, piezoresistive humidity sensors utilize a suspended structure coated with a water-absorbent layer. Polymer-based films have 2–6% water absorption, which results in a volume expansion ratio of approximately 5.5 × 10–5/%RH. Piezoresistors are used to detect the bending of this structure prompted by humidity changes. Buchhold et al. adopted the use of piezoresistors located on top of a diaphragm coated with a layer of polyimide. As the humidity increased, the humidity-dependent volume change of the polyimide layer prompted a deformation of the polyimide-bridge substrate bimorph and, therefore caused a bending of the plate, which was then transformed into an output voltage by an integrated piezoresistive bridge. It has been shown that the sensitivity of piezoresistive humidity sensors with resistors located at the periphery of the diaphragm does not depend significantly on the shape of the diaphragm. Hence, the specification of the diaphragm shape can be based solely on technological considerations.

Sager et al. investigated the humidity-dependent mechanical properties of polyimide films and their application to piezoresistive humidity sensors. The coefficient of humidity expansion (CHE) was shown to be approximately 60 to 80 ppm/%RH at different polyimide curing temperatures. The expansion can be used to create piezoresistive humidity sensors operating in a similar fashion to bimetal elements influenced by ambient temperature. The major difficulty in designing piezoresistive humidity sensors lies in providing an adequate thermal coupling to ensure a precise temperature control. Additionally, as in other sensor types, temperature drift is a significant problem for piezoresistive humidity sensors.

Magnetoelastic Humidity Sensors

The variation of the magnetoelastic resonance frequency could be used to monitor environmental parameters for remote query applications. It also could be used for relative humidity sensing. The transduction mecha-nism of magnetoelastic humidity sensors is based on the resonant frequency shift of the magnetic flux emitted from the sensors in response to a time varying magnitude field. For example, a remote query humidity sensor made by coating the magnetoelastic thick film with a thin layer of sol–gel deposited Al_2O_3 was reported by Grimes and Kouzoudis. In addition to humidity, other environmental parameters including temperature, DC magnetic field, surface stress, tension, liquid density and viscosity, displacement, atmospheric pressure, thin-film elastic modulus, level of liquids, mass loading and fluid-flow velocity could be monitored by the magnetoelastic sensors. Remote query measurement of chemicals could also be achieved using the type of magnetoelastic sensors, including glucose, NH_3, CO_2, and pH.

The operation principle of the magnetoelastic humidity sensors could be described as follows. The resonant frequency of the magnetoelastic sensor changes in response to mass loading within moisture sensing layers while relative humidity changes. Change of the mass results in a resonant frequency shift, which could be represented by the following formula:

$$\Delta f = -\frac{f}{2}\frac{\Delta m}{M}$$

Thus small changes in mass can be detected by monitoring the shift in the resonance frequency of the magnetoelastic sensor. The magnetoelastic sensors with a layer of moisture absorbing and desorbing material could be used for remote query measurement of humidity. The humidity-sensitive thin films including Al_2O_3, porous TiO_2, and $Fe_{73.5}Cu_1Nb_3Si_{13.5}B_9$ were reported in previous studies.

Biosensors

Biosensors can be defined as analytical devices which include a combination of biological detecting elements like sensor system and a transducer. When we compare with any other presently existing diagnostic device, these sensors are advanced in the conditions of selectivity as well as sensitivity. The applications of these Biosensors mainly include checking ecological pollution control, in agriculture field as well as food industries. The main features of biosensors are stability, cost, sensitivity, and reproducibility.

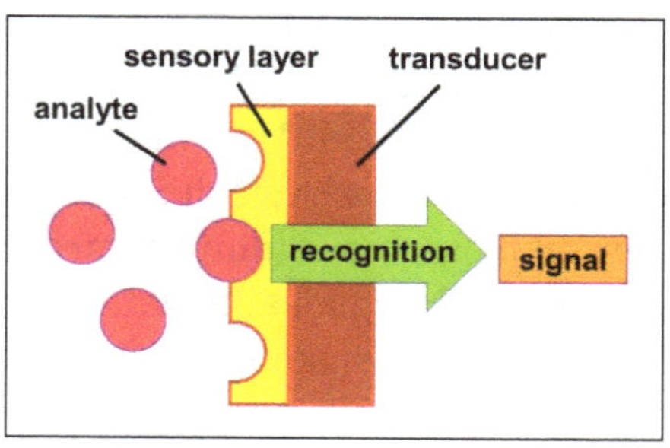

Main Components of a Biosensor

The block diagram of the biosensor includes three segments namely, sensor, transducer, and associated electrons. In the first segment, the sensor is a responsive biological part, the second segment is the detector part that changes the resulting signal from the contact of the analyte and for the results it displays in an accessible way. The final section comprises of an amplifier which is known as signal conditioning circuit, a display unit as well as the processor.

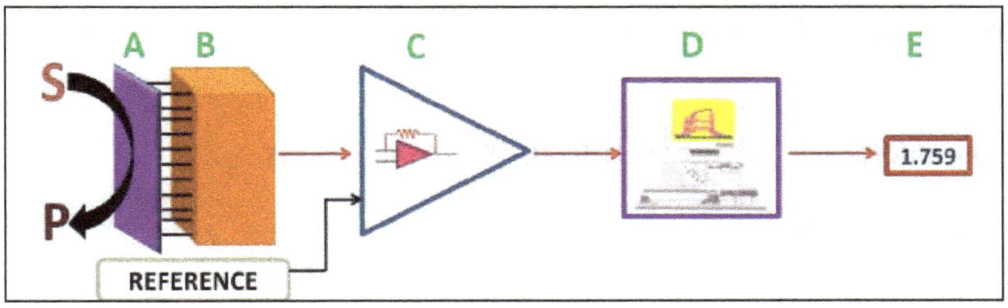

Working Principle of Biosensors

Usually, a specific enzyme or preferred biological material is deactivated by some of the usual methods, and the deactivated biological material is in near contact to the transducer. The analyte connects to the biological object to shape a clear analyte which in turn gives the electronic reaction that can be calculated. In some examples, the analyte is changed to a device which may be connected to the discharge of gas, heat, electron ions or hydrogen ions. In this, the transducer can alter the device linked converts into electrical signals which can be changed and calculated.

Working of Biosensors

The electrical signal of the transducer is frequently low and overlay upon a fairly high baseline. Generally, the signal processing includes deducting a position baseline signal, obtained from a related transducer without any biocatalyst covering.

The comparatively slow character of the biosensor reaction significantly eases the electrical noise filtration issue. In this stage, the direct output will be an analog signal however it is altered into digital form and accepted to a microprocessor phase where the information is progressed, influenced to preferred units and o/p to a data store.

Types of Biosensors

The different types of biosensors are classified based on the sensor device as well as the biological material that is discussed below.

Electrochemical Biosensor

Generally, the electrochemical biosensor is based on the reaction of enzymatic catalysis that consumes or generates electrons. Such types of enzymes are named as Redox Enzymes. The substrate of this biosensor generally includes three electrodes such as a counter, reference, and working type.

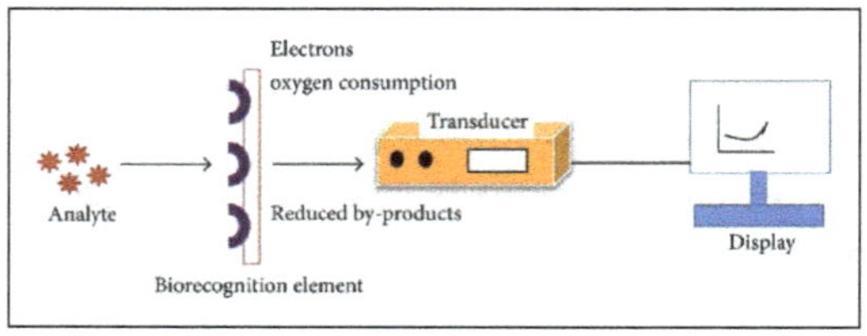

The object analyte is engaged in the response that happens on the surface of an active electrode, and this reaction may source also electron-transfer across the dual layer potential. The current can be calculated at a set potential.

Electrochemical biosensors are classified into four types:

- Amperometric Biosensors
- Potentiometric Biosensors
- Impedimetric Biosensors
- Voltammetric Biosensors

Amperometric Biosensor

An amperometric biosensor is a self-contained incorporated device based on the amount of the current ensuing from the oxidation offering exact quantitative analytical information.

Generally, these Biosensors have reaction times, energetic ranges & sensitivities comparable to the Potentiometric-biosensors. The simple amperometric biosensor in frequent usage includes the "Clark oxygen" electrode.

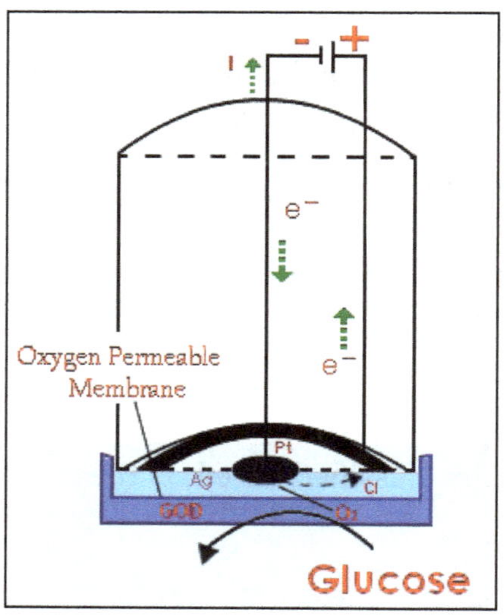

The rule of this biosensor is based on the amount of the flow of current between Counter Electrode and the working which is encouraged by a redox response at the operational electrode. Choosing analyte centers is essential for a wide selection of uses, comprising high-throughput medicine screening, quality control, problem finding and handling, and biological checking.

Potentiometric Biosensors

This type of biosensor provides a logarithmic reply by means of a high energetic range. These biosensors are frequently complete by monitor producing the electrode prototypes lying on a synthetic substrate, covered by a performing polymer with some enzyme is connected.

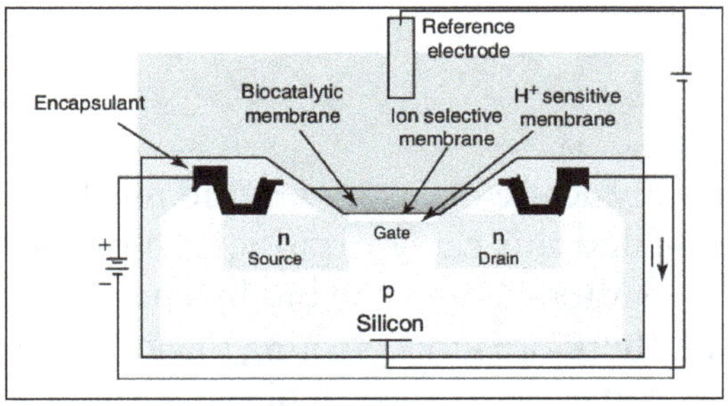

They comprise two electrodes which are enormously responsive and strong. They allow the recognition of analytes on stages before only attainable by HPLC, LC/MS & without exact model preparation.

All types of biosensors generally occupy least sample preparation because the biological detecting component is extremely choosy used for the analyte troubled. By the changes of physical and electrochemical the signal will be generated by in the layer of conducting polymer due to modifying happening at the outside of the biosensor.

These changes might be credited to ionic force, hydration, pH, and redox responses, the later as the label of enzyme rotating above a substrate. In FETs, the gate terminal has been changed with an antibody or enzyme, can also sense very-low attentions of different analytes because the required of analyte toward the gate terminal make a modify in the drain to source current.

Impedimetric Biosensors

The EIS (Electrochemical impedance spectroscopy) is a responsive indicator for a broad range of physical as well as chemical properties. A rising trend towards the expansion of Impedimetric-biosensors is being presently observed. The techniques of Impedimetric have been executed to differentiate the invention of the biosensors as well as to examine the catalyzed responses of enzymes lectins, nucleic acids, receptors, whole cells, and antibodies.

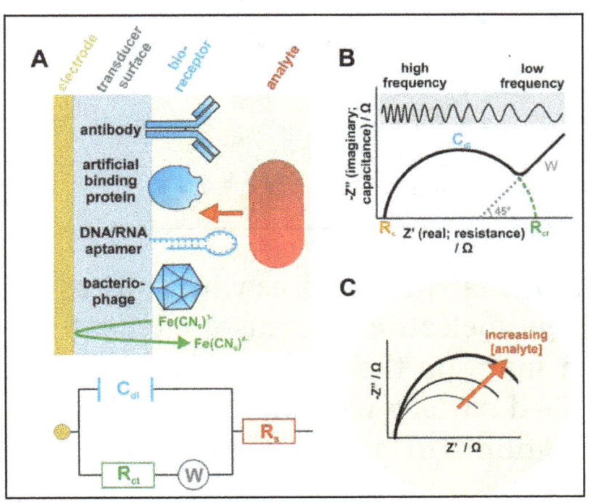

Voltammetric Biosensor

This communication is the base of a new voltammetric biosensor to notice acrylamide. This biosensor was built with a carbon glue electrode customized with Hb (hemoglobin), which includes four prostatic groups of the hem (Fe). This type of electrode shows a reversible oxidation or reduction procedure of Hb (Fe).

Physical Biosensor

In conditions of classification, physical biosensors are the most fundamental as well as broadly used sensors. The main ideas behind this categorization also happen from inspecting the human minds. As the general working method behind the intelligence of hearing, sight, touch is to react on the exterior physical stimuli, therefore any detecting device that offers reaction to the physical possessions of the medium was named as a physical biosensor.

The physical biosensors are classified into two types namely piezoelectric biosensor and thermometric biosensor.

Piezoelectric Biosensors

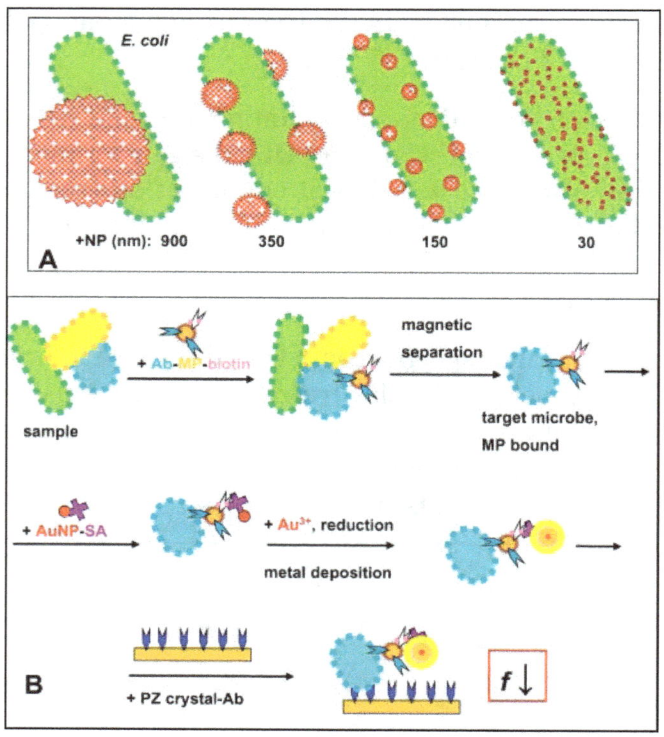

These sensors are a collection of analytical devices which works on a law of "affinity interaction recording". The platform of a piezoelectric is a sensor element works on the law of oscillations transform due to a collection jump on the surface of a piezoelectric crystal. In this analysis, biosensors having their modified surface with an antigen or antibody, a molecularly stamped polymer, and heritable information. The declared detection parts are normally united by using nanoparticles.

Thermometric Biosensor

There are various types of biological reactions which are connected with the invention of heat, and this makes the base of thermometric biosensors. These sensors are usually named as thermal biosensors.

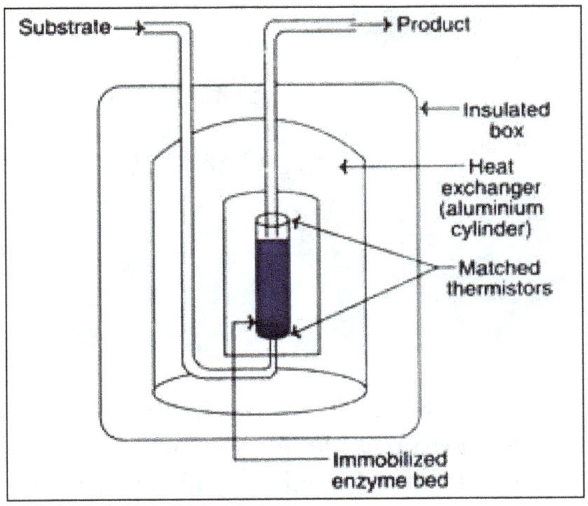

Thermometric-biosensor is used to measure or estimate the serum cholesterol. As cholesterol obtains oxidized through the enzyme cholesterol oxidize, then the heat will be produced which can be calculated. Similarly, assessments of glucose, urea, uric acid, and penicillin G can be done with these biosensors.

Optical Biosensor

The Optical biosensor is a device that uses an optical measurement principle. They use the fiber optics as well as optoelectronic transducers. The term optrode represents a compression of the two terms optical & electrode. These sensors mainly involve antibodies and enzymes like the transducing elements.

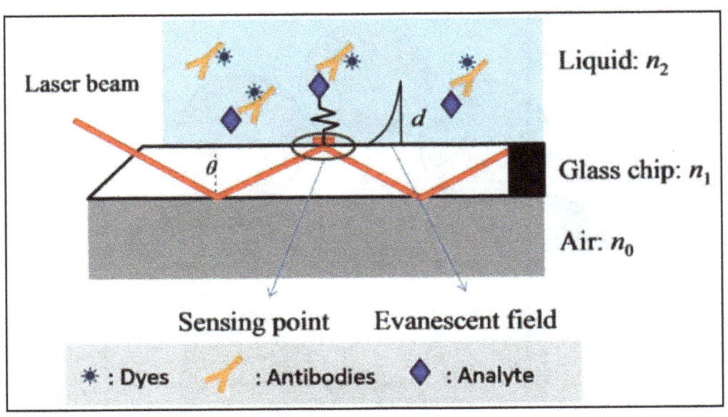

Optical biosensors permit a secure non-electrical inaccessible sensing of equipment. An extra benefit is that these frequently do not need reference sensors, because the comparative signal can be produced by using the similar light source like the sampling sensor. The optical biosensors are classified into two type's namely direct optical detection biosensor and labeled optical detection biosensor.

Wearable Biosensors

The wearable biosensor is a digital device, used to wear on the human body in different wearable systems like smart watches, smart shirts, tattoos which allows the levels of blood glucose, BP, the rate of heartbeat, etc.

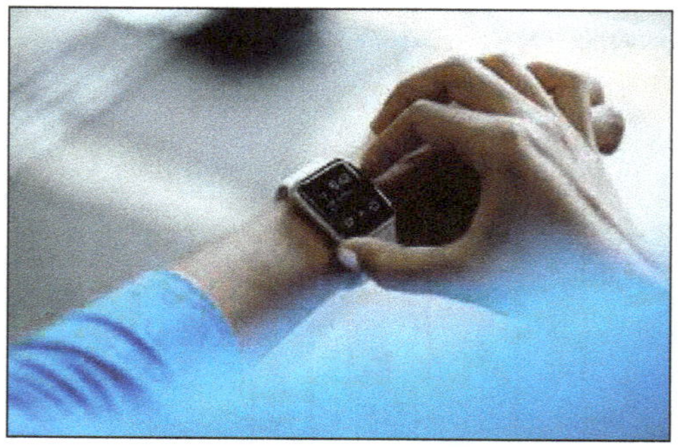

Nowadays, we can notice that these sensors are carrying out a signal of improvement to the world. Their better use and ease can give an original level of experience into a patient's real-time fitness status. This data accessibility will let superior clinical choice and will effect in enhanced health results and extra capable use of health systems.

For the human beings, these sensors may assist in premature recognition of health actions and prevention of hospitalization. The possibility of these sensors to reduce hospital stays and readmissions will definitely attract positive awareness in the upcoming future. As well, investigate information says that WBS will definitely carry a cost-effective wearable health equipment to the world.

Biosensors Applications

In recent years, these sensors have become very popular, and they are applicable in different fields which are mentioned below.

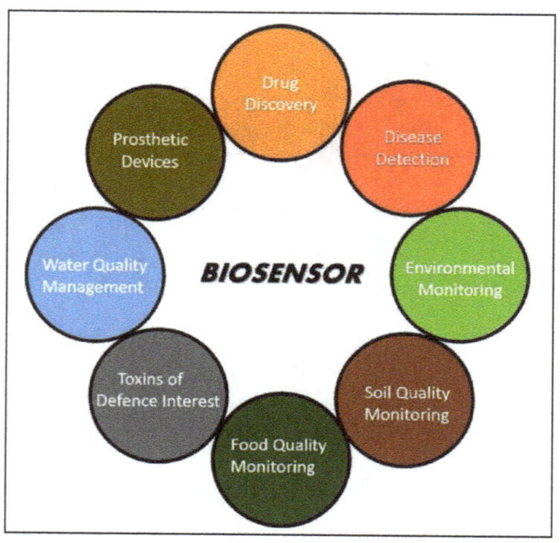

- Common healthcare checking.
- Metabolites Measurement.

- Screening for sickness.
- Insulin treatment.
- Clinical psychotherapy & diagnosis of disease.
- In Military.
- Agricultural, and Veterinary applications.
- Drug improvement, offense detection.
- Processing & monitoring in Industrial.
- Ecological pollution control.

Optical Sensors

An optical sensor is a device that converts light rays into electronic signals. Similar to a photo resistor, it measures the physical quantity of light and translates it into a form read by the instrument. One of the features of an optical sensor is its ability to measure the changes from one or more light beams. This change is most often based around alterations to the intensity of the light. Optical sensors can work either on the single point method or through a distribution of points. Through the single point method, a sole phase change is needed to activate the sensor. In terms of the distribution concept, the sensor is reactive along a long series of sensors or single fiberoptic array.

Other features of optical sensors include the distinction of whether it is placed internally or externally in a device.

Today Optical Sensors are used in numerous research, and commercial applications such as for quality and process control, medico technologies, metrology, imaging, and remote sensing. There are many types of optical sensors; many based on the use of lasers, imaging systems, and/or fibers. In addition, novel sensor methods that enable more advanced sensing are continuously being developed by using novel materials, such as meta materials, micro and nano structured materials or by employing new frequency bands as for example THz radiation.

Advantages and Disadvantages of Optical Sensors

Research and development in the optical sensor field is motivated by the expectation that optical sensors have significant advantages compared to conventional sensor types, in terms of their properties. Below is given list of some of the advantages of optical over non optical sensors:

- Greater sensitivity.
- Electrical Passiveness.

- Freedom from Electromagnetic interference.
- Wide dynamic range.
- Both points and distributed configuration.
- Multiplexing capabilities.

Table: Comparison of Extrinsic and Intrinsic optical sensors.

Extrinsic	Intrinsic
• Applications - Temperature, pressure, liquid level and flow. • Less sensitive • Easily multiplexed • Ingress/ egress connection problems • Easier to use • Less expensive	• Applications - Rotation, acceleration, • Strain, acoustic pressure and vibration. • More sensitive • Tougher to multiplex • Reduces connection problems • More elaborate signal demodulation • More expensive

Taking advantage of the capacity of optical fibers to send and receive optical signals over long distances, a current trend is to create networks of sensors, or sensor arrays. This avoids having to convert between electronics and photonics separately at each sensing site, thereby reducing costs and increasing flexibility. A difficulty of all sensors, both optical and non optical, is interference from multiple effects. A sensor intended to measure strain or pressure may be very temperature-sensitive.

Optical Sensor Technologies

Measurands and Sensor Categories

Through evolution of optical sensing technology, one can measure nearly all of the physical measurands of interest and a very large number of chemical quantities. The measurands possible are listed below:

- Temperature
- Pressure
- Flow
- Liquid level
- Displacement (Position)
- Vibration
- Rotation
- Magnetic fields
- Acceleration
- Chemical species
- Force
- Radiation
- Ph
- Humidity
- Strain
- Velocity
- Electric fields
- Acoustic field

Means of Sensing

In this category, sensors are generally based either on measuring an intensity change in one or more light beams or on looking at phase changes in the light beams by causing them to interact or interfere with one another. Thus sensors in this category are termed either intensity sensors or interferometric sensors. Techniques used in the case of intensity sensors include light scattering (both Rayleigh and Raman), spectral transmission changes (i.e., simple attenuation of transmitted light due to absorption), micro bending or radioactive losses, reflectance changes, and changes in the modal properties of the fiber.

Extent of Sensing

This category is based on whether sensors operate only at a single point or over a distribution of points. Thus, sensors in this category are termed either point sensors or distributed sensors. In the case of a point sensor, the transducer may be at the end of a fiber. Examples of this sensor type are fiber Bragg gratings distributed along a fiber length to measure strain or temperature.

Types of Optical Sensors

Optical sensor has two points. One is the transmitting point where light is emitted and the other end is the receiving end. Generally there are 3 types of optical sensor, through beam, reflective and retro reflective. Each of this type has different advantages.

Through Beam Sensor

This sensor is suitable for absolute detection of solid object. The transmitter and receiver are pointing to each other directly in order to create a straight light path. Whenever there is any object which passes by the light transmission path, the amount of received light by the receiver will be zero or reduced and the sensing method is interrupting the light source received by the receiver.

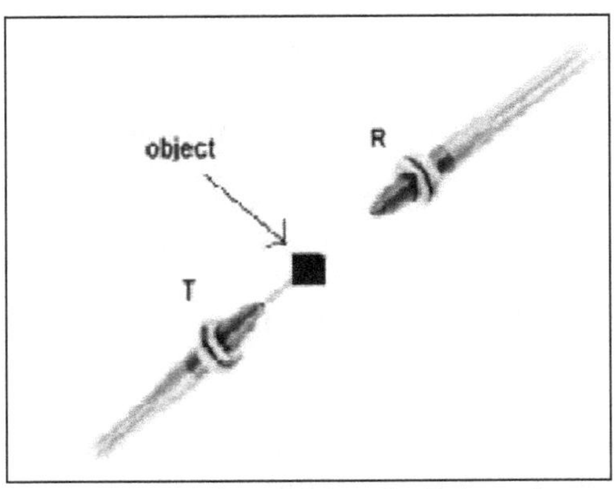

With object interupting the light path.

Reflective Sensor: This type of sensor if suitable to perform color differentiation. The transmitter and receiver of this sensor are parallel to each other. The method of detection is through reflection of light emitted by the receiver.

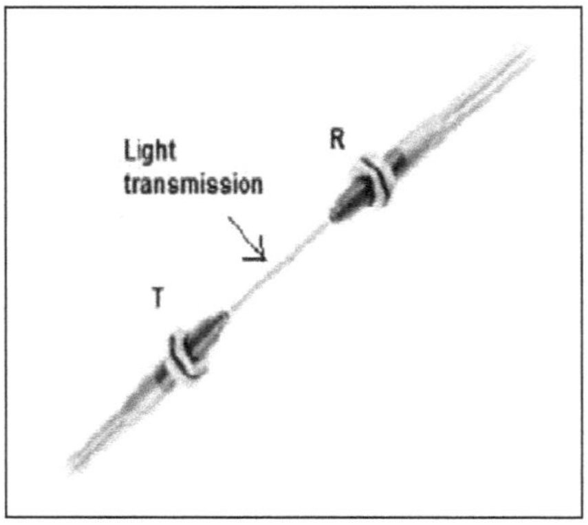

Without object.

If we are to use reflective sensor, we should take note that this sensor actually does detection base on quantity of light that is reflected back to the receiver. If the light source used is red LED, we will face problem to differentiate red and white objects. Red and white actually reflect back almost the same amount of light.

Advantages of Optical Sensors

1. Completely passive: Can be used in explosive environment.
2. Immune to electromagnetic interference: Ideal for microwave environment.
3. Resistant to high temperatures and chemically reactive environment: Ideal for harsh and hostile environment.
4. Small size: Ideal for embedding and surface mounting.
5. High degree of biocompatibility, non-intrusive nature and electromagnetic immune: Ideal-for medical applications like intra-aortic balloon pumping.
6. Can monitor a wide range of physical and chemical parameters.
7. Potential for very high sensitivity, range and resolution.
8. Complete electrical insulation from high electrostatic potential.
9. Remote operation over several km lengths without any lead sensitivity: Ideal for deployment in boreholes or measurements in hazardous environment.
10. Multiplexed and distributed sensors are unique in that they provide measurements at a large number of points along a single.
11. Fiber optic sensors are excellent candidates for monitoring environmental changes and they offer many advantages over conventional electronic sensors as listed below:
    - Easy integration into a wide variety of structures, including composite materials, with little interference due to their small size and cylindrical geometry.
    - Inability to conduct electric current.

- Immune to electromagnetic interference and radio frequency interference.
- Lightweight.
- Robust, more resistant to harsh environments.
- High sensitivity.
- Multiplexing capability to form sensing networks.
- Remote sensing capability.
- Multifunctional sensing capabilities such as strain, pressure, corrosion, temperature and acoustic signals.

12. Techniques by which the measurements are made can be broadly grouped in three categories depending on:
    - How the sensing is accomplished.
    - The physical extent of the sensing.
    - The role of the optical fiber in the sensing process.

Means of Sensing

In this category, sensors are generally based either on measuring an intensity change in one or more light beams or on looking at phase changes in the light beams by causing them to interact or interfere with one another. Thus sensors in this category are termed either intensity sensors or interferometric sensors. Techniques used in the case of intensity sensors include light scattering (both Rayleigh and Raman), spectral transmission changes (i.e., simple attenuation of transmitted light due to absorption), micro bending or radioactive losses, reflectance changes, and changes in the modal properties of the fiber. Interferometric sensors have been demonstrated based upon the magneto-optic, the laser-Doppler, or the Signac effects.

Extent of Sensing

This category is based on whether sensors operate only at a single point or over a distribution of points. Thus, sensors in this category are termed either point sensors or distributed sensors. Point sensor: detect measure and variation only in the vicinity of the sensor.

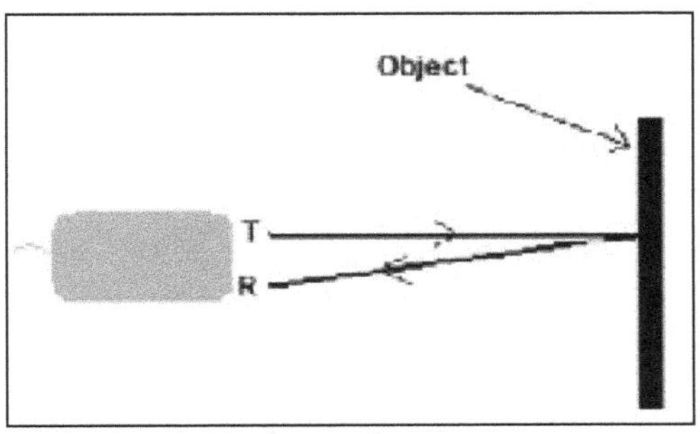

Reflective Sensor.

Multiplexed Sensor

Multiple localized sensors are placed at intervals along the fiber length.

Distributed Sensor

Sensing is distributed along the length of the fiber.

Optical Fiber

An optical fiber is a thin, flexible, transparent fiber that acts as a waveguide, or "light pipe", to transmit light between the two ends of the fiber. The field of applied science and engineering concerned with the design and application of optical fibers is known as fiber optics. Fibers are used instead of metal wires because signals travel along them with less loss and are also immune to electromagnetic interference. Specially designed fibers are used for a variety of other applications, including sensors and fiber lasers.

Optical fiber typically consists of a transparent core surrounded by a transparent cladding material with a lower index of refraction. Light is kept in the core by total internal reflection. This causes the fiber to act as a waveguide. Fibers which support many propagation paths or transverse modes are called multimode fibers (MMF), while those which can only support a single mode are called single-mode fibers (SMF). Multimode fibers generally have a larger core diameter, and are used for short-distance communication links and for applications where high power must be transmitted. Single-mode fibers are used for most communication links longer than 1,050 meters (3,440 ft).

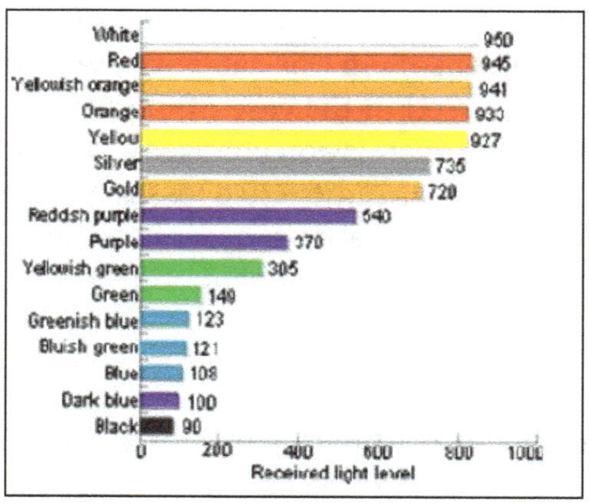

Colour differentiation chart.

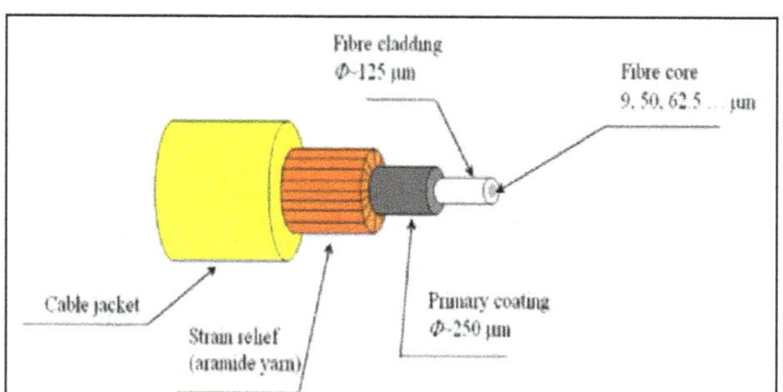

Sensors and fibre lasers.

Figures above show a schematic of a step-index optical fiber. Light is guided inside the core region by total internal reflection at the corecladding interface. Depending on the size of the core region, one single or multiple light paths (modes) are permitted to propagate, referred to as single-mode or multimode fiber. Typically, the bare optical fiber has an outer diameter of 125μm with a core diameter of 9μm in the case of single-mode fibers and 50μm or 62.5μm for multimode fibers Different protective coatings are applied to protect the fiber from possible mechanical damage.

Role of Optical Fiber

Extrinsic sensors are those where the light leaves the feed or transmitting fiber to be changed before it continues to the detector by means of the return or receiving fiber. Intrinsic sensors are different in that the light beam doesn't leave the optical fiber but is changed whilst still contained within it.

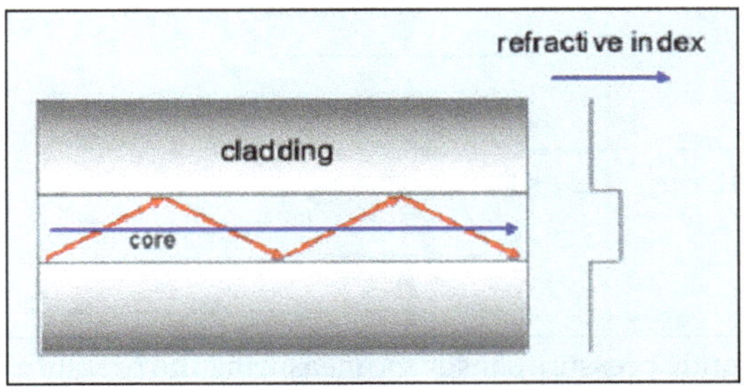

Schematic of step-index optical fibre.

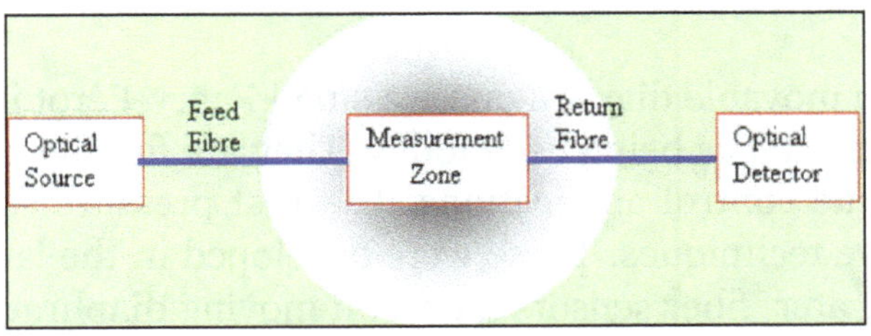

The simple sensor detects any increase or decrease in the length/between the two fibers. As the distance between the two is increased the amount of light launched into the return fiber will decrease. Conversely as the length is decreased the light intensity collected by the receiver will increase giving a relatively simple fiber optic sensor for determining small shifts between objects. While a sensor of this kind has problems with sensitivity to lateral movement it is a good illustration of a basic sensing technique.

As the pressure increases the fiber will become slightly deformed and experience increased micro bending losses which results in a decrease in the light intensity received at the detector. A decrease in the pressure relieves stress on the fibre and hence there is an increase in transmitted light detected.

Sensor Types

As a result of the myriad ways now available to sense the same quantity, no single sensing technique has emerged to become the large-volume leader. Some techniques, however, seem to be more prominent than others for sensing a given measurand, and each technique tends to have its own specialists among the company and university labs. This is definitely a field in which new technologies are being developed and tested continuously; it is this plethora of new techniques that leads to the fragmented nature of the optical sensor marketplace.

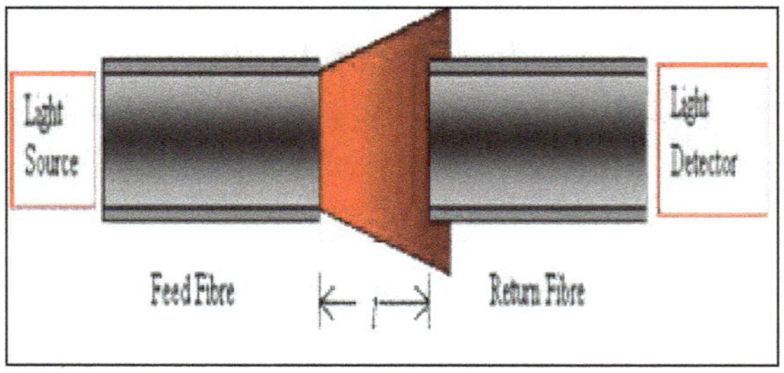

An intensity based extrinsic sensor for measuring the distance l between two objects.

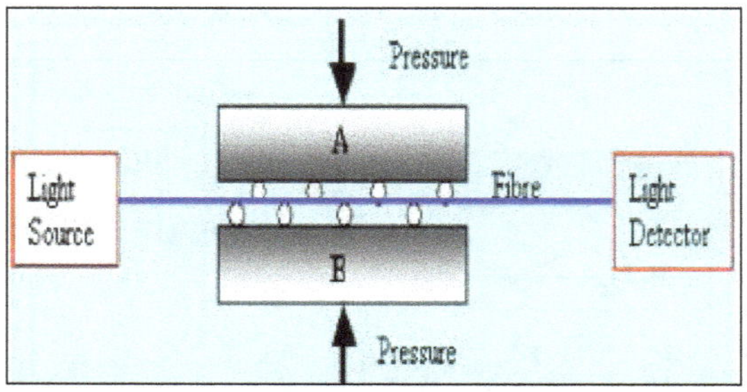

An intensity based fibre optic pressure sensor for measuring the pressure between plates A and B.

Pressure Sensor

Pressure sensors based on movable diaphragms, on small Fabry-Pérot interferometers, or on micro bending, are the primary types being used today. They are finding use in biomedical, process control, marine, and engine control applications. The first pressure sensors for biomedical usage relied on piezoresistive techniques. These were developed in the late 1950s for intravascular pressure measurements. Later, fiber sensors based on moving diaphragms and monitoring retro reflected intensity emerged. Camino Labs in San Diego, CA, manufactures devices of this type and is reported to be producing around 60,000 devices/year. Consider the schematic of an optical pressure transducer shown in figure below.

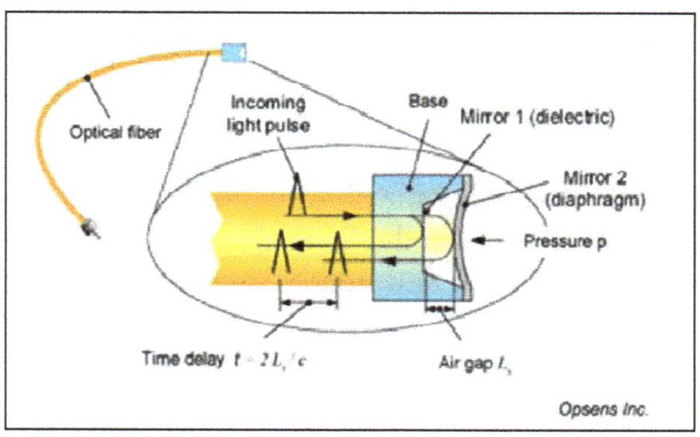

Schematic of pressure transducer based on Fabry-Perot cavity.

Essentially, it consists of a pair of parallel mirrors separated by an air gap Ls. This arrangement is referred to as a Fabry-Perot (FP) cavity or sensing interferometer. A semi-reflective mirror 1 is formed by depositing a dielectric layer at the end of the optical fiber. Mirror 2 is formed by a diaphragm mounted in front of the optical fiber. Exposing the diaphragm to the pressure p to be measured changes the gap Ls. Hence, by measuring Ls the applied pressure p can be determined. Different pressure ranges can be accommodated by appropriately selecting thickness and diameter of the diaphragm to keep the maximum deflection of similar value and maintain a linear relation between pressure and deflection.

Flow Sensor

Flow sensor is the use of pressure sensors in conjunction with the venture effect to measure flow. Differential pressure is measured between two segments of a venture tube that have a different aperture. The pressure difference between the two segments is directly proportional to the flow rate through the venture tube:

$$F= \text{sqrt}\ (p1\text{-}p2)$$

A pressure sensor may be used to sense the decay of pressure due to a system leak. This is commonly done by means of utilizing the pressure sensor to measure pressure change over time.

Displacement and Position Sensors

Displacement sensors were some of the first optoelectronic sensors to be developed, beginning in the late 1970s and early 1980s. These simplest sensors rely on the change in retro reflectance of light into a fiber because of movement of a proximal mirror surface. One of the first Photometric sensors was of this type, in which a conical tip was applied to the end of a fiber. In this light is totally reflected back into the fiber if the surrounding medium is air; however, if the fiber is inserted into a liquid matching the fiber index, light is extracted from the fiber and lost. Thus, displacement of the liquid surface can be tracked. For obvious reasons, these displacement sensors are referred to as liquid level sensors.

This technique is commonly employed to measure the depth of a submerged body (such as a diver or submarine), or level of contents in a tank (such as in a water tower). For most practical purposes, fluid level is directly proportional to pressure. The basic equation for such a measurement is:

$$P= pgh$$

where, P = pressure, p = density of the fluid, g = standard gravity, h = height of fluid column above pressure sensor.

Temperature Sensors

Temperature sensors probably constitute the largest class of commercially available optical sensors. Many different physical phenomena are used to perform the sensing, each with attributes suitable for a particular application; no single technique can accommodate the entire range of temperatures and resolutions required for different applications.

Optical fiber based distributed sensors have been widely used to monitor temperature. The main advantage of the system is that the fiber itself is the sensing element. Distributed temperature sensing technology shows real advantages over conventional temperature sensing technology when a temperature profile of the installation is required or when a large number of sensing points is crucial. Therefore this technology lends itself to long length applications (pipelines, tunnels, power cables, conveyor belts), applications where only small sensors can access (oil wells) and safety critical applications where it is important to have all points monitored (refineries, LNG plants, electrochemical processes).

In Raman scattering based distributed temperature sensor a pulsed laser is injected into the optical fiber which is the sensing element. In the fiber the photons interact with the molecules of the fiber material. The spectrum of the backscattered light includes the Rayleigh, the Brillion and the Raman backscattered light. The Raman backscattered light is caused by thermally influenced molecular vibrations. Consequently, the Raman backscattered light carries the information on the temperature of the fiber and can be used to obtain information about the temperature distribution along the fiber. The Raman backscattering light has two components: the Stokes (I) and the Anti-Stokes (I) component. They can be separated from the primary and the Rayleigh backscattered light due to their differences in wavelength.

The Stokes component is only weakly dependent on temperature, while the Anti-Stokes component shows a strong relation to temperature. The ratio of the intensities of Stokes and Anti-Stokes components is a measure of temperature. Since the injected light is a pulse of a few nanoseconds, the time of arrival of instantaneous back scattered intensity can be correlated with the distance along the fiber length from where it is scattered.

Applications

Industrial Applications

A variety of in-process inspection and control functions can be performed through non-contact optical sensors. Here are some examples of applications being developed at IMI:

1. Temperature measurement by infrared sensors.
2. Surface and subsurface detection of defects and delaminations.
3. Surface inspection for gauging roughness and thickness.
4. Characterization of particle flows (temperature, velocity and diameter).
5. In-process composition analysis by laser-induced plasma spectroscopy.
6. Real-time thermal imaging of dynamic processes.
7. In-situ monitoring of coatings and lubricant films.

Biomedical Applications

Flow monitoring by laser Dopplerimetry is used in several biomedical sensing applications, including dermatology for testing of skin irritants, gastroenterology via endoscopes for making blood perfusion measurements in the stomach and duodenum, etc., and dentistry for contact probes to measure blood flow in the teeth and gums. Sensors are used in internal medicine for angiology and vascular surgery to monitor blood flow during vascular reconstruction and the degree of arteriosclerosis in arteries, and they are used in orthopedics for monitoring the blood perfusion in tissues during and after surgery.

Fibre optic sensors have proven to be ideal transducers for structural monitoring. Being durable, stable and insensitive to external perturbations, they are particularly interesting for the long-term health assessment of civil and geotechnical structures. Many different fibre optic sensor technologies exist and offer a wide range of performances and suitability for different applications. The most widely used sensing techniques include point sensors (Fibre Bragg Gratings and Fabry-Perot interferometers), long-gauge sensors (SOFO) and distributed sensors (Raman and Brillouin scattering sensors). These sensing technologies are now widely used in routine application for health monitoring of structures such as bridges, buildings, monuments, tunnels, dams, dykes, pipelines, landslides and many others. This contribution reviews these systems and technologies and presents some significant application examples, in particular to bridges, buildings, geostructures and pipelines.

Biometrics Applications

Optical sensor scans our finger, it functions like a camera, and obtains our fingerprint image indirectly. We may have similar kind of experience: when a government agency takes our fingerprint, someone will check our fingerprint to see if it is visible. Then, they will put some lotion on our finger, and the operator will press our finger on scanner to take our fingerprint. Optical sensor works in a similar way, so it needs good visible finger, clean finger surface, not too dry and so on.

Applications of Fiber Optic Sensors

Fiber optic sensors are used in several areas. Specifically:

- Measurement of physical properties such as strain, displacement, temperature, pressure, velocity, and acceleration in structures of any shape or size.
- Monitoring the physical health of structures in real time.
- Buildings and Bridges: Concrete monitoring during setting, crack (length, propagation speed) monitoring, prestressing monitoring, spatial displacement measurement, neutral axis evolution, long-term deformation (creep and shrinkage) monitoring, concrete-steel interaction, and post-seismic damage evaluation.
- Tunnels: Multipoint optical extensometers, convergence monitoring, concrete/prefabricated vaults evaluation, and joints monitoring damage detection.
- Dams: Foundation monitoring, joint expansion monitoring, spatial displacement measurement, leakage monitoring, and distributed temperature monitoring.
- Heritage structures: Displacement monitoring, crack opening analysis, post-seismic damage evaluation, restoration monitoring, and old-new interaction.

Position Sensors

Position Sensors detect the position of something which means that they are referenced either to or from some fixed point or position. These types of sensors provide a "positional" feedback.

One method of determining a position, is to use either "distance", which could be the distance between two points such as the distance travelled or moved away from some fixed point, or by "rotation" (angular movement). For example, the rotation of a robots wheel to determine its distance travelled along the ground. Either way, Position Sensors can detect the movement of an object in a straight line using Linear Sensors or by its angular movement using Rotational Sensors.

Potentiometer

The most commonly used of all the "Position Sensors", is the potentiometer because it is an inexpensive and easy to use position sensor. It has a wiper contact linked to a mechanical shaft that can be either angular (rotational) or linear (slider type) in its movement, and which causes the resistance value between the wiper/slider and the two end connections to change giving an electrical signal output that has a proportional relationship between the actual wiper position on the resistive track and its resistance value. In other words, resistance is proportional to position.

Potentiometer

Potentiometers come in a wide range of designs and sizes such as the commonly available round rotational type or the longer and flat linear slider types. When used as a position sensor the moveable object is connected directly to the rotational shaft or slider of the potentiometer.

A DC reference voltage is applied across the two outer fixed connections forming the resistive element. The output voltage signal is taken from the wiper terminal of the sliding contact.

This configuration produces a potential or voltage divider type circuit output which is proportional to the shaft position. Then for example, if you apply a voltage of say 10v across the resistive element of the potentiometer the maximum output voltage would be equal to the supply voltage at 10 volts, with the minimum output voltage equal to 0 volts. Then the potentiometer wiper will vary the output signal from 0 to 10 volts, with 5 volts indicating that the wiper or slider is at its half-way or centre position.

Potentiometer Construction

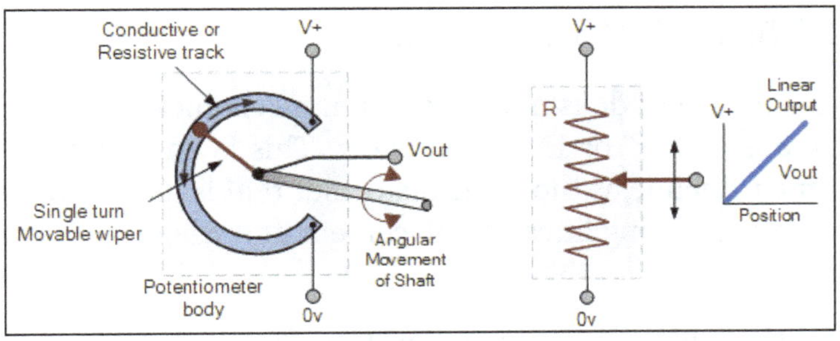

The output signal (Vout) from the potentiometer is taken from the centre wiper connection as it moves along the resistive track, and is proportional to the angular position of the shaft.

Example of a Simple Positional Sensing Circuit

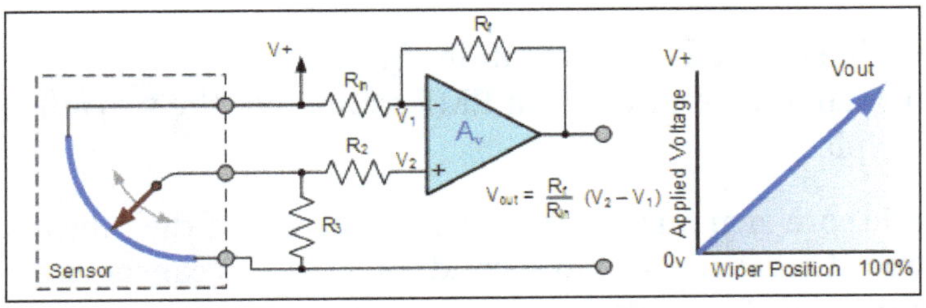

While resistive potentiometer position sensors have many advantages: low cost, low tech, easy to use etc, as a position sensor they also have many disadvantages: wear due to moving parts, low accuracy, low repeatability, and limited frequency response.

But there is one main disadvantage of using the potentiometer as a positional sensor. The range of movement of its wiper or slider (and hence the output signal obtained) is limited to the physical size of the potentiometer being used.

For example a single turn rotational potentiometer generally only has a fixed mechanical rotation of between 0o and about 240 to 330o maximum. However, multi-turn pots of up to 3600o (10 x 360o) of mechanical rotation are also available.

Most types of potentiometers use carbon film for their resistive track, but these types are electrically noisy (the crackle on a radio volume control), and also have a short mechanical life.

Wire-wound pots also known as rheostats, in the form of either a straight wire or wound coil resistive wire can also be used, but wire wound pots suffer from resolution problems as their wiper jumps from one wire segment to the next producing a logarithmic (LOG) output resulting in errors in the output signal. These too suffer from electrical noise.

For high precision low noise applications conductive plastic resistance element type polymer film or cermet type potentiometers are now available. These pots have a smooth low friction electrically linear (LIN) resistive track giving them a low noise, long life and excellent resolution and are available as both multi-turn and single turn devices. Typical applications for this type of high accuracy position sensor is in computer game joysticks, steering wheels, industrial and robot applications.

Inductive Position Sensors

Linear Variable Differential Transformer

One type of positional sensor that does not suffer from mechanical wear problems is the "Linear Variable Differential Transformer" or LVDT for short. This is an inductive type position sensor which works on the same principle as the AC transformer that is used to measure movement. It is a very accurate device for measuring linear displacement and whose output is proportional to the position of its moveable core.

It basically consists of three coils wound on a hollow tube former, one forming the primary coil and the other two coils forming identical secondaries connected electrically together in series but 1800 out of phase either side of the primary coil.

A moveable soft iron ferromagnetic core (sometimes called an "armature") which is connected to the object being measured, slides or moves up and down inside the tubular body of the LVDT.

A small AC reference voltage called the "excitation signal" (2 – 20V rms, 2 – 20kHz) is applied to the primary winding which in turn induces an EMF signal into the two adjacent secondary windings (transformer principles).

If the soft iron magnetic core armature is exactly in the centre of the tube and the windings, "null position", the two induced emf's in the two secondary windings cancel each other out as they are 1800 out of phase, so the resultant output voltage is zero. As the core is displaced slightly to one side or the other from this null or zero position, the induced voltage in one of the secondaries will be become greater than that of the other secondary and an output will be produced.

The polarity of the output signal depends upon the direction and displacement of the moving core. The greater the movement of the soft iron core from its central null position the greater will be the resulting output signal. The result is a differential voltage output which varies linearly with the cores position. Therefore, the output signal from this type of position sensor has both an amplitude that is a linear function of the cores displacement and a polarity that indicates direction of movement.

The phase of the output signal can be compared to the primary coil excitation phase enabling suitable electronic circuits such as the AD592 LVDT Sensor Amplifier to know which half of the coil the magnetic core is in and thereby know the direction of travel.

The Linear Variable Differential Transformer

When the armature is moved from one end to the other through the centre position the output voltages changes from maximum to zero and back to maximum again but in the process changes its phase angle by 180 deg's. This enables the LVDT to produce an output AC signal whose magnitude represents the amount of movement from the centre position and whose phase angle represents the direction of movement of the core.

A typical application of a linear variable differential transformer (LVDT) sensor would be as a pressure transducer, were the pressure being measured pushes against a diaphragm to produce a force. The force is then converted into a readable voltage signal by the sensor.

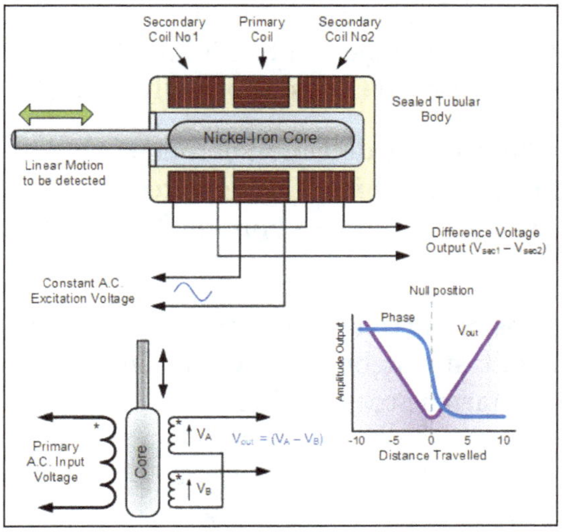

Advantages of the linear variable differential transformer, or LVDT compared to a resistive potentiometer are that its linearity, that is its voltage output to displacement is excellent, very good accuracy, good resolution, high sensitivity as well as frictionless operation. They are also sealed for use in hostile environments.

Inductive Proximity Sensors

Another type of inductive position sensor in common use is the Inductive Proximity Sensor also called an Eddy current sensor. While they do not actually measure displacement or angular rotation they are mainly used to detect the presence of an object in front of them or within a close proximity, hence their name "proximity sensor".

Proximity sensors, are non-contact position sensors that use a magnetic field for detection with the simplest magnetic sensor being the reed switch. In an inductive sensor, a coil is wound around an iron core within an electromagnetic field to form an inductive loop.

When a ferromagnetic material is placed within the eddy current field generated around the inductive sensor, such as a ferromagnetic metal plate or metal screw, the inductance of the coil changes significantly. The proximity sensors detection circuit detects this change producing an output voltage. Therefore, inductive proximity sensors operate under the electrical principle of Faraday's Law of inductance.

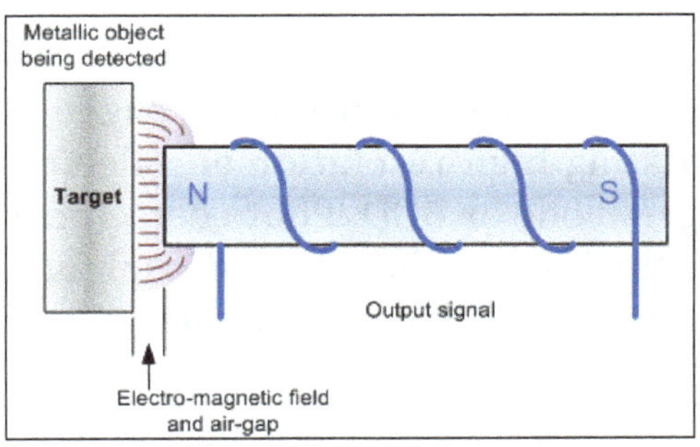

An inductive proximity sensor has four main components; The oscillator which produces the electromagnetic field, the coil which generates the magnetic field, the detection circuit which detects any change in the field when an object enters it and the output circuit which produces the output signal, either with normally closed (NC) or normally open (NO) contacts.

Inductive proximity sensors allow for the detection of metallic objects in front of the sensor head without any physical contact of the object itself being detected. This makes them ideal for use in dirty or wet environments. The "sensing" range of proximity sensors is very small, typically 0.1mm to 12mm.

As well as industrial applications, inductive proximity sensors are also commonly used to control the flow of traffic by changing of traffic lights at junctions and cross roads. Rectangular inductive loops of wire are buried into the tarmac road surface.

Proximity Sensor

When a car or other road vehicle passes over this inductive loop, the metallic body of the vehicle changes the loops inductance and activates the sensor thereby alerting the traffic lights controller that there is a vehicle waiting.

One main disadvantage of these types of position sensors is that they are "Omni-directional", that is they will sense a metallic object either above, below or to the side of it. Also, they do not detect non-metallic objects although Capacitive Proximity Sensors and Ultrasonic Proximity Sensors are available. Other commonly available magnetic positional sensors include: reed switches, Hall Effect Sensors and variable reluctance sensors.

Rotary Encoders

Rotary Encoders are another type of position sensor which resemble potentiometers mentioned earlier but are non-contact optical devices used for converting the angular position of a rotating shaft into an analogue or digital data code. In other words, they convert mechanical movement into an electrical signal (preferably digital).

All optical encoders work on the same basic principle. Light from an LED or infrared light source is passed through a rotating high-resolution encoded disk that contains the required code patterns, either binary, grey code or BCD. Photo detectors scan the disk as it rotates and an electronic circuit processes the information into a digital form as a stream of binary output pulses that are fed to counters or controllers which determine the actual angular position of the shaft.

There are two basic types of rotary optical encoders, Incremental Encoders and Absolute Position Encoders.

Incremental Encoder

Incremental Encoders, also known as quadrature encoders or relative rotary encoder, are the simplest of the two position sensors. Their output is a series of square wave pulses generated by a photocell arrangement as the coded disk, with evenly spaced transparent and dark lines called segments on its surface, moves or rotates past the light source. The encoder produces a stream of square wave pulses which, when counted, indicates the angular position of the rotating shaft.

Incremental encoders have two separate outputs called "quadrature outputs". These two outputs are displaced at 900 out of phase from each other with the direction of rotation of the shaft being determined from the output sequence.

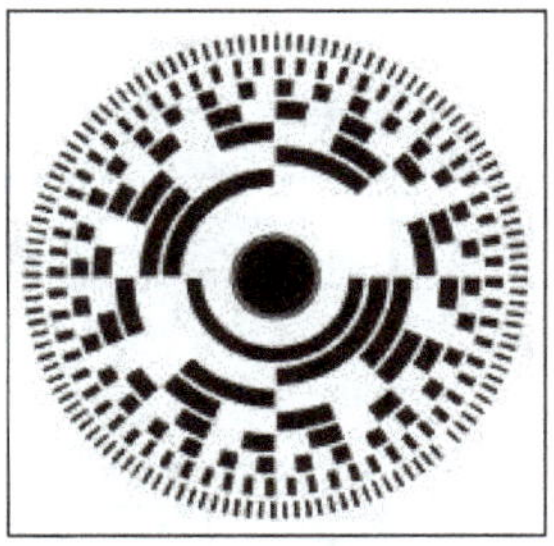

Encoder Disk

The number of transparent and dark segments or slots on the disk determines the resolution of the device and increasing the number of lines in the pattern increases the resolution per degree of rotation. Typical encoded discs have a resolution of up to 256 pulses or 8-bits per rotation.

The simplest incremental encoder is called a tachometer. It has one single square wave output and is often used in unidirectional applications where basic position or speed information only is required. The "Quadrature" or "Sine wave" encoder is the more common and has two output square waves commonly called channel A and channel B. This device uses two photo detectors, slightly offset from each other by 900 thereby producing two separate sine and cosine output signals.

Simple Incremental Encoder

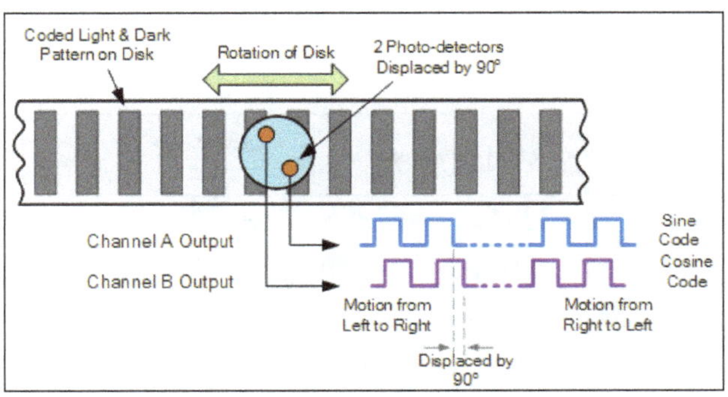

By using the Arc Tangent mathematical function the angle of the shaft in radians can be calculated. Generally, the optical disk used in rotary position encoders is circular, then the resolution of the output will be given as: $\theta = 360/n$, where n equals the number of segments on coded disk.

Then for example, the number of segments required to give an incremental encoder a resolution of

10 will be: 1° = 360/n, therefore, n = 360 windows, etc. Also the direction of rotation is determined by noting which channel produces an output first, either channel A or channel B giving two directions of rotation, A leads B or B leads A.

Incremental Encoder Output

One main disadvantage of incremental encoders when used as a position sensor, is that they require external counters to determine the absolute angle of the shaft within a given rotation. If the power is momentarily shut off, or if the encoder misses a pulse due to noise or a dirty disc, the resulting angular information will produce an error. One way of overcoming this disadvantage is to use absolute position encoders.

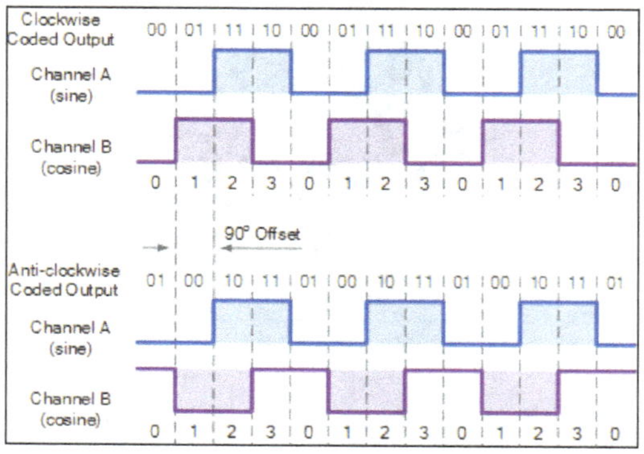

Absolute Position Encoder

Absolute Position Encoders are more complex than quadrature encoders. They provide a unique output code for every single position of rotation indicating both position and direction. Their coded disk consists of multiple concentric "tracks" of light and dark segments. Each track is independent with its own photo detector to simultaneously read a unique coded position value for each angle of movement. The number of tracks on the disk corresponds to the binary "bit"-resolution of the encoder so a 12-bit absolute encoder would have 12 tracks and the same coded value only appears once per revolution.

4-bit Binary Coded Disc

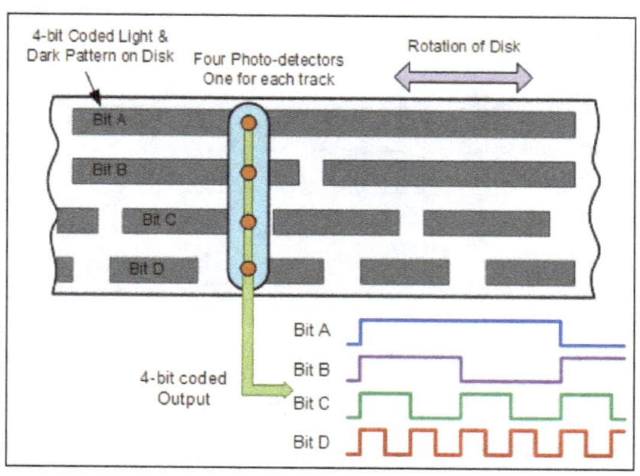

One main advantage of an absolute encoder is its non-volatile memory which retains the exact position of the encoder without the need to return to a "home" position if the power fails. Most rotary encoders are defined as "single-turn" devices, but absolute multi-turn devices are available, which obtain feedback over several revolutions by adding extra code disks.

Typical application of absolute position encoders are in computer hard drives and CD/DVD drives were the absolute position of the drives read/write heads are monitored or in printers/plotters to accurately position the printing heads over the paper.

Light Sensors

A Light Sensor generates an output signal indicating the intensity of light by measuring the radiant energy that exists in a very narrow range of frequencies basically called "light", and which ranges in frequency from "Infrared" to "Visible" up to "Ultraviolet" light spectrum.

The light sensor is a passive devices that convert this "light energy" whether visible or in the infrared parts of the spectrum into an electrical signal output. Light sensors are more commonly known as "Photoelectric Devices" or "Photo Sensors" because the convert light energy (photons) into electricity (electrons).

Photoelectric devices can be grouped into two main categories, those which generate electricity when illuminated, such as Photo-voltaics or Photo-emissives etc, and those which change their electrical properties in some way such as Photo-resistors or Photo-conductors. This leads to the following classification of devices.

- Photo-emissive Cells – These are photodevices which release free electrons from a light sensitive material such as caesium when struck by a photon of sufficient energy. The amount of energy the photons have depends on the frequency of the light and the higher the frequency, the more energy the photons have converting light energy into electrical energy.
- Photo-conductive Cells – These photodevices vary their electrical resistance when subjected to light. Photoconductivity results from light hitting a semiconductor material which controls the current flow through it. Thus, more light increase the current for a given applied voltage. The most common photoconductive material is Cadmium Sulphide used in LDR photocells.
- Photo-voltaic Cells – These photodevices generate an emf in proportion to the radiant light energy received and is similar in effect to photoconductivity. Light energy falls on to two semiconductor materials sandwiched together creating a voltage of approximately 0.5V. The most common photovoltaic material is Selenium used in solar cells.
- Photo-junction Devices – These photodevices are mainly true semiconductor devices such as the photodiode or phototransistor which use light to control the flow of electrons and holes across their PN-junction. Photojunction devices are specifically designed for detector application and light penetration with their spectral response tuned to the wavelength of incident light.

Photoconductive Cell

A Photoconductive light sensor does not produce electricity but simply changes its physical properties when subjected to light energy. The most common type of photoconductive device is the Photoresistor which changes its electrical resistance in response to changes in the light intensity.

Photoresistors are Semiconductor devices that use light energy to control the flow of electrons, and hence the current flowing through them. The commonly used Photoconductive Cell is called the Light Dependent Resistor or LDR.

Light Dependent Resistor

As its name implies, the Light Dependent Resistor (LDR) is made from a piece of exposed semiconductor material such as cadmium sulphide that changes its electrical resistance from several thousand Ohms in the dark to only a few hundred Ohms when light falls upon it by creating hole-electron pairs in the material.

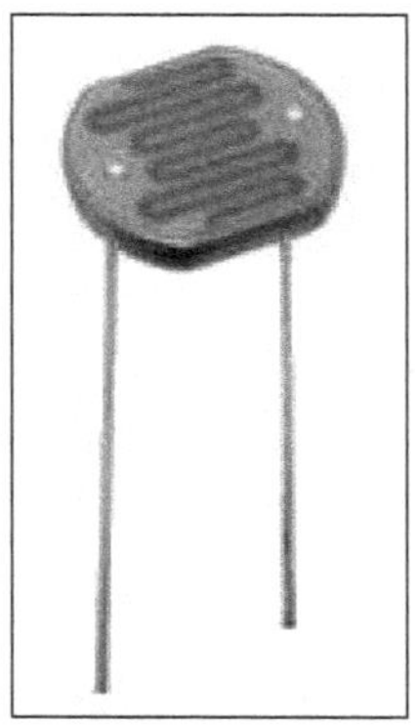

Typical LDR

The net effect is an improvement in its conductivity with a decrease in resistance for an increase in illumination. Also, photoresistive cells have a long response time requiring many seconds to respond to a change in the light intensity.

Materials used as the semiconductor substrate include, lead sulphide (PbS), lead selenide (PbSe), indium antimonide (InSb) which detect light in the infrared range with the most commonly used of all photoresistive light sensors being Cadmium Sulphide (Cds).

Cadmium sulphide is used in the manufacture of photoconductive cells because its spectral response curve closely matches that of the human eye and can even be controlled using a simple torch as a light source. Typically then, it has a peak sensitivity wavelength (λp) of about 560nm to 600nm in the visible spectral range.

Light Dependent Resistor Cell

The most commonly used photoresistive light sensor is the ORP12 Cadmium Sulphide photoconductive cell. This light dependent resistor has a spectral response of about 610nm in the yellow to orange region of light. The resistance of the cell when unilluminated (dark resistance) is very high at about $10M\Omega$'s which falls to about $100M\Omega$'s when fully illuminated (lit resistance).

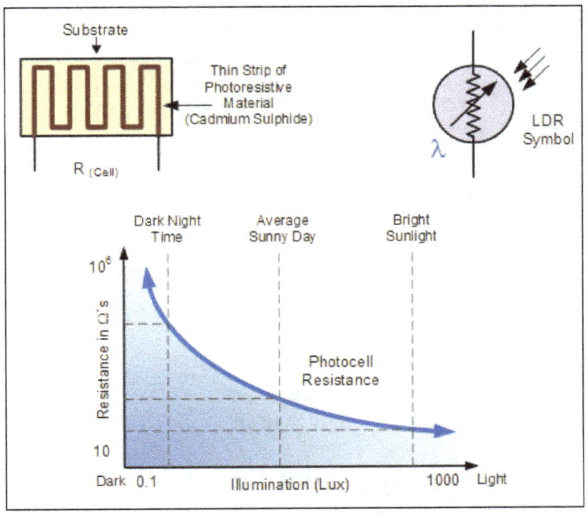

To increase the dark resistance and therefore reduce the dark current, the resistive path forms a zigzag pattern across the ceramic substrate. The CdS photocell is a very low cost device often used in auto dimming, darkness or twilight detection for turning the street lights "ON" and "OFF", and for photographic exposure meter type applications.

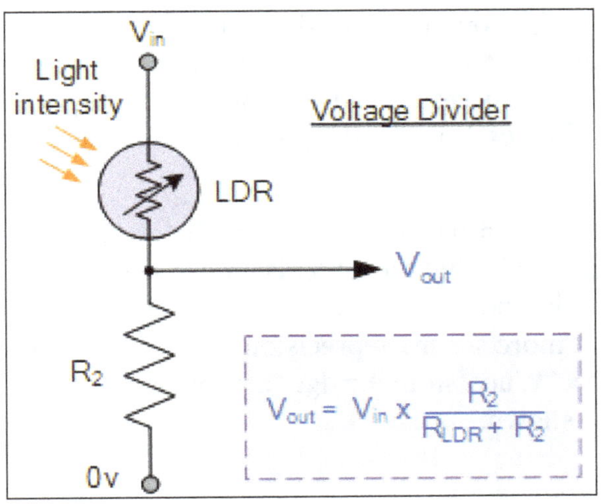

Connecting a light dependant resistor in series with a standard resistor like this across a single DC supply voltage has one major advantage, a different voltage will appear at their junction for different levels of light.

The amount of voltage drop across series resistor, R_2 is determined by the resistive value of the light dependant resistor, R_{LDR}. This ability to generate different voltages produces a very handy circuit called a "Potential Divider" or Voltage Divider Network.

As we know, the current through a series circuit is common and as the LDR changes its resistive value due to the light intensity, the voltage present at V_{OUT} will be determined by the voltage divider formula. An LDR's resistance, R_{LDR} can vary from about 100 Ω in the sun light, to over 10 MΩ in absolute darkness with this variation of resistance being converted into a voltage variation at VOUT as shown.

One simple use of a Light Dependent Resistor, is as a light sensitive switch as shown below.

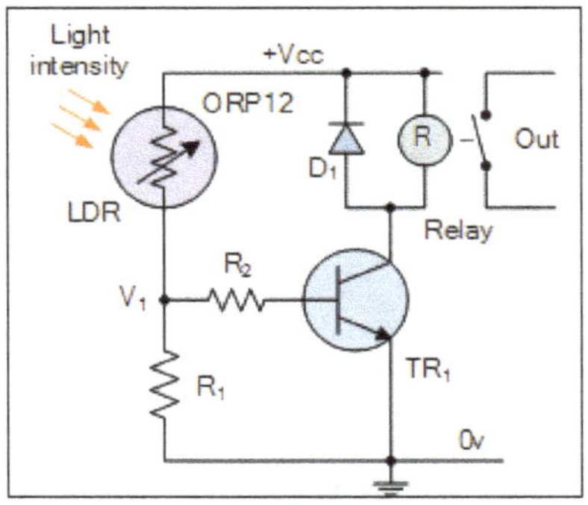

LDR Switch

This basic light sensor circuit is of a relay output light activated switch. A potential divider circuit is formed between the photoresistor, LDR and the resistor R_1. When no light is present ie in darkness, the resistance of the LDR is very high in the Megaohms ($M\Omega$) range so zero base bias is applied to the transistor TR_1 and the relay is de-energised or "off".

As the light level increases the resistance of the LDR starts to decrease causing the base bias voltage at V_1 to rise. At some point determined by the potential divider network formed with resistor R_1, the base bias voltage is high enough to turn the transistor TR_1 "on" and thus activate the relay which in turn is used to control some external circuitry. As the light level falls back to darkness again the resistance of the LDR increases causing the base voltage of the transistor to decrease, turning the transistor and relay "off" at a fixed light level determined again by the potential divider network.

By replacing the fixed resistor R1 with a potentiometer VR1, the point at which the relay turns "on" or "off" can be pre-set to a particular light level. This type of simple circuit shown above has a fairly low sensitivity and its switching point may not be consistent due to variations in either temperature or the supply voltage. A more sensitive precision light activated circuit can be easily made by incorporating the LDR into a "Wheatstone Bridge" arrangement and replacing the transistor with an Operational Amplifier as shown.

Light Level Sensing Circuit

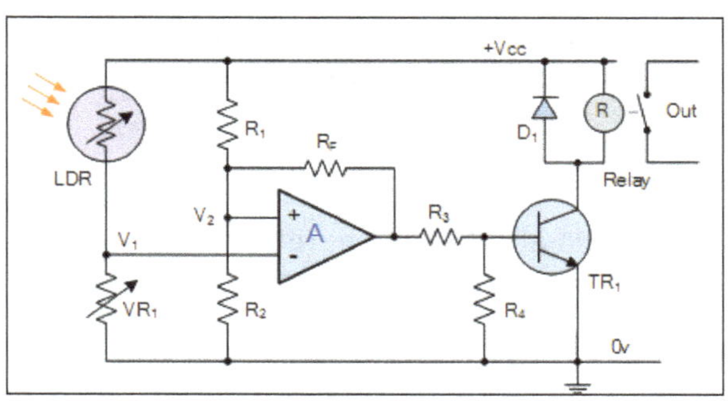

In this basic dark sensing circuit, the light dependent resistor LDR_1 and the potentiometer VR_1 form one adjustable arm of a simple resistance bridge network, also known commonly as a Wheatstone bridge, while the two fixed resistors R_1 and R_2 form the other arm. Both sides of the bridge form potential divider networks across the supply voltage whose outputs V_1 and V_2 are connected to the non-inverting and inverting voltage inputs respectively of the operational amplifier.

The operational amplifier is configured as a Differential Amplifier also known as a voltage comparator with feedback whose output voltage condition is determined by the difference between the two input signals or voltages, V_1 and V_2. The resistor combination R_1 and R_2 form a fixed voltage reference at input V_2, set by the ratio of the two resistors. The LDR – VR_1 combination provides a variable voltage input V_1 proportional to the light level being detected by the photoresistor.

As with the previous circuit the output from the operational amplifier is used to control a relay, which is protected by a free wheel diode, D_1. When the light level sensed by the LDR and its output voltage falls below the reference voltage set at V_2 the output from the op-amp changes state activating the relay and switching the connected load.

Likewise as the light level increases the output will switch back turning “off” the relay. The hysteresis of the two switching points is set by the feedback resistor Rf can be chosen to give any suitable voltage gain of the amplifier.

The operation of this type of light sensor circuit can also be reversed to switch the relay “on” when the light level exceeds the reference voltage level and vice versa by reversing the positions of the light sensor LDR and the potentiometer VR_1. The potentiometer can be used to “pre-set” the switching point of the differential amplifier to any particular light level making it ideal as a simple light sensor project circuit.

Photojunction Devices

Photojunction Devices are basically PN-Junction light sensors or detectors made from silicon semiconductor PN-junctions which are sensitive to light and which can detect both visible light and infrared light levels. Photo-junction devices are specifically made for sensing light and this class of photoelectric light sensors include the Photodiode and the Phototransistor.

The Photodiode

The construction of the Photodiode light sensor is similar to that of a conventional PN-junction diode except that the diodes outer casing is either transparent or has a clear lens to focus the light onto the PN junction for increased sensitivity. The junction will respond to light particularly longer wavelengths such as red and infrared rather than visible light.

Photo-diode

This characteristic can be a problem for diodes with transparent or glass bead bodies such as the 1N4148 signal diode. LED's can also be used as photodiodes as they can both emit and detect light from their junction. All PN-junctions are light sensitive and can be used in a photo-conductive unbiased voltage mode with the PN-junction of the photodiode always "Reverse Biased" so that only the diodes leakage or dark current can flow.

The current-voltage characteristic (I/V Curves) of a photodiode with no light on its junction (dark mode) is very similar to a normal signal or rectifying diode. When the photodiode is forward biased, there is an exponential increase in the current, the same as for a normal diode. When a reverse bias is applied, a small reverse saturation current appears which causes an increase of the depletion region, which is the sensitive part of the junction. Photodiodes can also be connected in a current mode using a fixed bias voltage across the junction. The current mode is very linear over a wide range.

Photo-diode Construction and Characteristics

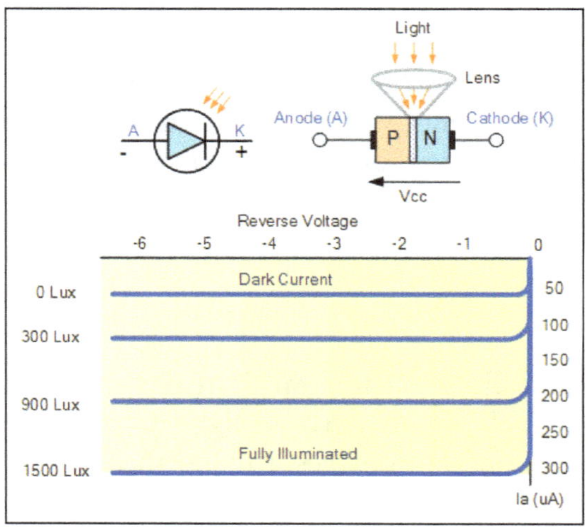

When used as a light sensor, a photodiodes dark current (0 lux) is about 10uA for geranium and 1uA for silicon type diodes. When light falls upon the junction more hole/electron pairs are formed and the leakage current increases. This leakage current increases as the illumination of the junction increases.

Thus, the photodiodes current is directly proportional to light intensity falling onto the PN-junction. One main advantage of photodiodes when used as light sensors is their fast response to changes in the light levels, but one disadvantage of this type of photodevice is the relatively small current flow even when fully lit.

The following circuit shows a photo-current-to-voltage converter circuit using an operational amplifier as the amplifying device. The output voltage (Vout) is given as $Vout = I_P * Rf$ and which is proportional to the light intensity characteristics of the photodiode.

This type of circuit also utilizes the characteristics of an operational amplifier with two input terminals at about zero voltage to operate the photodiode without bias. This zero-bias op-amp configuration gives a high impedance loading to the photodiode resulting in less influence by dark current and a wider linear range of the photocurrent relative to the radiant light intensity. Capacitor C_f is used to prevent oscillation or gain peaking and to set the output bandwidth $(1/2\pi RC)$.

Photo-diode Amplifier Circuit

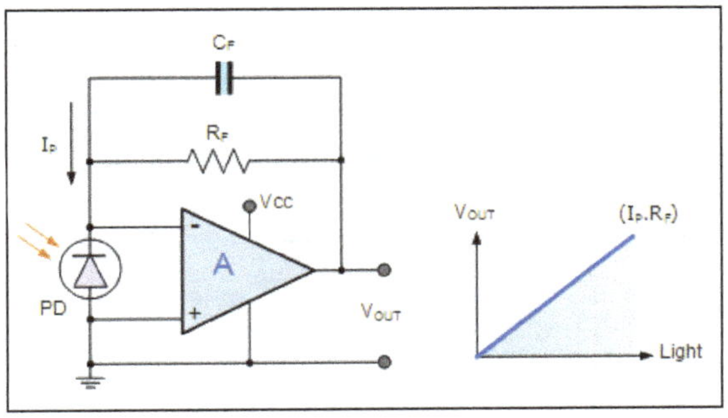

Photodiodes are very versatile light sensors that can turn its current flow both "on" and "off" in nanoseconds and are commonly used in cameras, light meters, CD and DVD-ROM drives, TV remote controls, scanners, fax machines and copiers etc, and when integrated into operational amplifier circuits as infrared spectrum detectors for fibre optic communications, burglar alarm motion detection circuits and numerous imaging, laser scanning and positioning systems etc.

Phototransistor

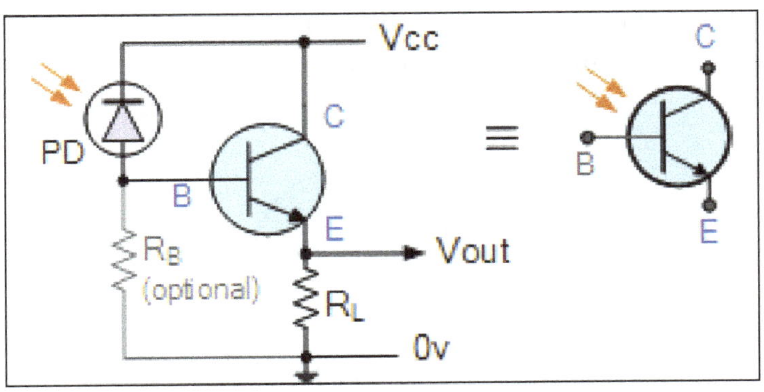

An alternative photo-junction device to the photodiode is the Phototransistor which is basically a photodiode with amplification. The Phototransistor light sensor has its collector-base PN-junction reverse biased exposing it to the radiant light source.

Phototransistors operate the same as the photodiode except that they can provide current gain and are much more sensitive than the photodiode with currents are 50 to 100 times greater than that of the standard photodiode and any normal transistor can be easily converted into a phototransistor light sensor by connecting a photodiode between the collector and base.

Phototransistors consist mainly of a bipolar NPN Transistor with its large base region electrically unconnected, although some phototransistors allow a base connection to control the sensitivity, and which uses photons of light to generate a base current which in turn causes a collector to emitter current to flow. Most phototransistors are NPN types whose outer casing is either transparent or has a clear lens to focus the light onto the base junction for increased sensitivity.

Photo-transistor Construction and Characteristics

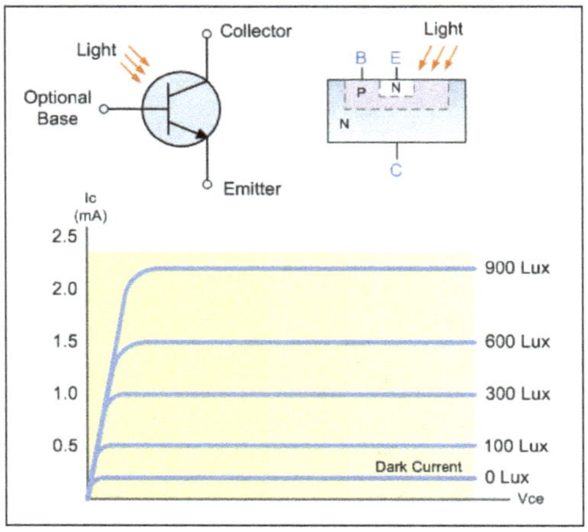

In the NPN transistor the collector is biased positively with respect to the emitter so that the base/collector junction is reverse biased. therefore, with no light on the junction normal leakage or dark current flows which is very small. When light falls on the base more electron/hole pairs are formed in this region and the current produced by this action is amplified by the transistor.

Usually the sensitivity of a phototransistor is a function of the DC current gain of the transistor. Therefore, the overall sensitivity is a function of collector current and can be controlled by connecting a resistance between the base and the emitter but for very high sensitivity optocoupler type applications, Darlington phototransistors are generally used.

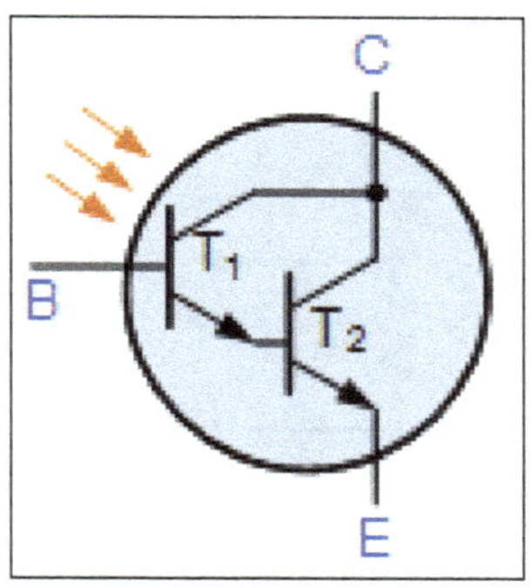

Photo-darlington

Photodarlington transistors use a second bipolar NPN transistor to provide additional amplification or when higher sensitivity of a photodetector is required due to low light levels or selective sensitivity, but its response is slower than that of an ordinary NPN phototransistor.

Photo darlington devices consist of a normal phototransistor whose emitter output is coupled to the base of a larger bipolar NPN transistor. Because a darlington transistor configuration gives a current gain equal to a product of the current gains of two individual transistors, a photodarlington device produces a very sensitive detector.

Typical applications of Phototransistors light sensors are in opto-isolators, slotted opto switches, light beam sensors, fibre optics and TV type remote controls, etc. Infrared filters are sometimes required when detecting visible light.

Another type of photojunction semiconductor light sensor worth a mention is the Photo-thyristor. This is a light activated thyristor or Silicon Controlled Rectifier, SCR that can be used as a light activated switch in AC applications. However their sensitivity is usually very low compared to equivalent photodiodes or phototransistors.

To help increase their sensitivity to light, photo-thyristors are made thinner around the gate junction. The downside to this process is that it limits the amount of anode current that they can switch. Then for higher current AC applications they are used as pilot devices in opto-couplers to switch larger more conventional thyristors.

Photovoltaic Cells

The most common type of photovoltaic light sensor is the Solar Cell. Solar cells convert light energy directly into DC electrical energy in the form of a voltage or current to a power a resistive load such as a light, battery or motor. Then photovoltaic cells are similar in many ways to a battery because they supply DC power.

However, unlike the other photo devices we have looked at above which use light intensity even from a torch to operate, photovoltaic solar cells work best using the suns radiant energy.

Solar cells are used in many different types of applications to offer an alternative power source from conventional batteries, such as in calculators, satellites and now in homes offering a form of renewable power.

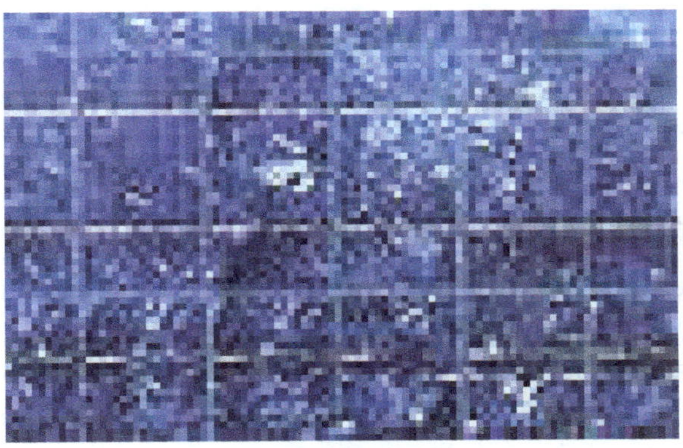

Photovoltaic Cell

Photovoltaic cells are made from single crystal silicon PN junctions, the same as photodiodes with a very large light sensitive region but are used without the reverse bias. They have the same characteristics as a very large photodiode when in the dark.

When illuminated the light energy causes electrons to flow through the PN junction and an individual solar cell can generate an open circuit voltage of about 0.58v (580mV). Solar cells have a "Positive" and a "Negative" side just like a battery.

Individual solar cells can be connected together in series to form solar panels which increases the output voltage or connected together in parallel to increase the available current. Commercially available solar panels are rated in Watts, which is the product of the output voltage and current (Volts times Amps) when fully lit.

Characteristics of a Typical Photovoltaic Solar Cell

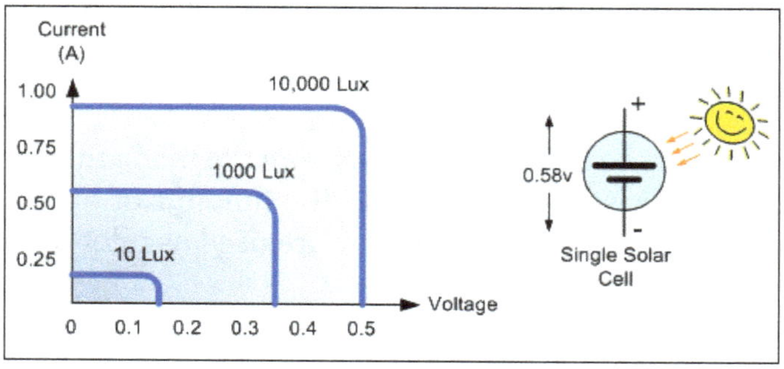

The amount of available current from a solar cell depends upon the light intensity, the size of the cell and its efficiency which is generally very low at around 15 to 20%. To increase the overall efficiency of the cell commercially available solar cells use polycrystalline silicon or amorphous silicon, which have no crystalline structure, and can generate currents of between 20 to 40mA per cm^2.

Other materials used in the construction of photovoltaic cells include Gallium Arsenide, Copper Indium Diselenide and Cadmium Telluride. These different materials each have a different spectrum band response, and so can be "tuned" to produce an output voltage at different wavelengths of light.

Hall Effect Sensor

Magnetic sensors convert magnetic or magnetically encoded information into electrical signals for processing by electronic circuits.

Magnetic sensors are solid state devices that are becoming more and more popular because they can be used in many different types of application such as sensing position, velocity or directional movement. They are also a popular choice of sensor for the electronics designer due to their non-contact wear free operation, their low maintenance, robust design and as sealed hall effect devices are immune to vibration, dust and water.

One of the main uses of magnetic sensors is in automotive systems for the sensing of position, distance and speed. For example, the angular position of the crank shaft for the firing angle of the spark plugs, the position of the car seats and seat belts for air-bag control or wheel speed detection for the anti-lock braking system, (ABS).

Magnetic sensors are designed to respond to a wide range of positive and negative magnetic fields in a variety of different applications and one type of magnet sensor whose output signal is a function of magnetic field density around it is called the Hall Effect Sensor.

Hall Effect Sensors are devices which are activated by an external magnetic field. We know that a magnetic field has two important characteristics flux density, (B) and polarity (North and South Poles). The output signal from a Hall effect sensor is the function of magnetic field density around the device. When the magnetic flux density around the sensor exceeds a certain pre-set threshold, the sensor detects it and generates an output voltage called the Hall Voltage, V_H. Consider the diagram below.

Hall Effect Sensor Principles

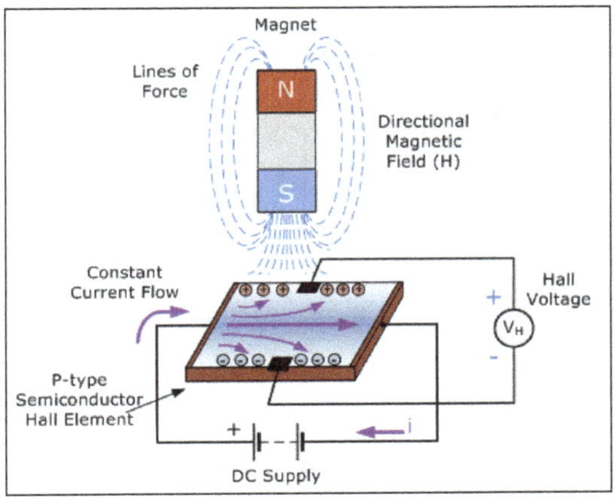

Hall Effect Sensors consist basically of a thin piece of rectangular p-type semiconductor material such as gallium arsenide (GaAs), indium antimonide (InSb) or indium arsenide (InAs) passing a continuous current through itself. When the device is placed within a magnetic field, the magnetic flux lines exert a force on the semiconductor material which deflects the charge carriers, electrons and holes, to either side of the semiconductor slab. This movement of charge carriers is a result of the magnetic force they experience passing through the semiconductor material.

As these electrons and holes move side wards a potential difference is produced between the two sides of the semiconductor material by the build-up of these charge carriers. Then the movement of electrons through the semiconductor material is affected by the presence of an external magnetic field which is at right angles to it and this effect is greater in a flat rectangular shaped material.

The effect of generating a measurable voltage by using a magnetic field is called the Hall Effect after Edwin Hall who discovered it back in the 1870's with the basic physical principle underlying the Hall effect being Lorentz force. To generate a potential difference across the device the magnetic flux lines must be perpendicular, (900) to the flow of current and be of the correct polarity, generally a south pole.

The Hall effect provides information regarding the type of magnetic pole and magnitude of the magnetic field. For example, a south pole would cause the device to produce a voltage output while a north pole would have no effect. Generally, Hall Effect sensors and switches are designed to be in the "off", (open circuit condition) when there is no magnetic field present. They only turn "on", (closed circuit condition) when subjected to a magnetic field of sufficient strength and polarity.

Hall Effect Magnetic Sensor

The output voltage, called the Hall voltage, (V_H) of the basic Hall Element is directly proportional to the strength of the magnetic field passing through the semiconductor material (output α H). This output voltage can be quite small, only a few microvolts even when subjected to strong magnetic fields so most commercially available Hall effect devices are manufactured with built-in DC amplifiers, logic switching circuits and voltage regulators to improve the sensors sensitivity, hysteresis and output voltage. This also allows the Hall effect sensor to operate over a wider range of power supplies and magnetic field conditions.

Hall Effect Sensor

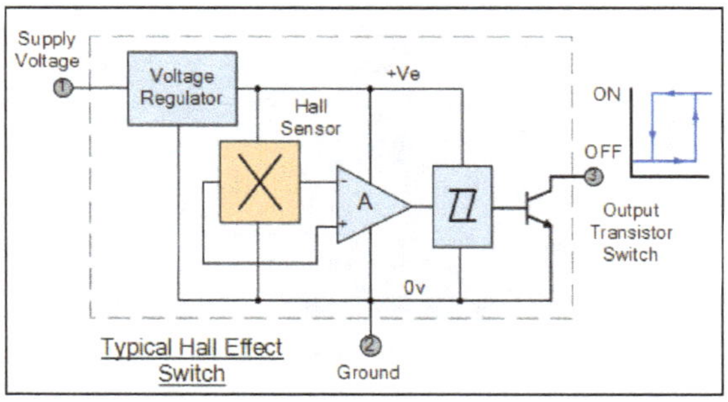

Hall Effect Sensors are available with either linear or digital outputs. The output signal for linear (analogue) sensors is taken directly from the output of the operational amplifier with the output voltage being directly proportional to the magnetic field passing through the Hall sensor. This output Hall voltage is given as:

$$V_H = R_H \left(\frac{I}{t} \times B \right)$$

where:

- V_H is the Hall Voltage in volts.
- R_H is the Hall Effect co-efficient.
- I is the current flow through the sensor in amps.
- t is the thickness of the sensor in mm.
- B is the Magnetic Flux density in Teslas.

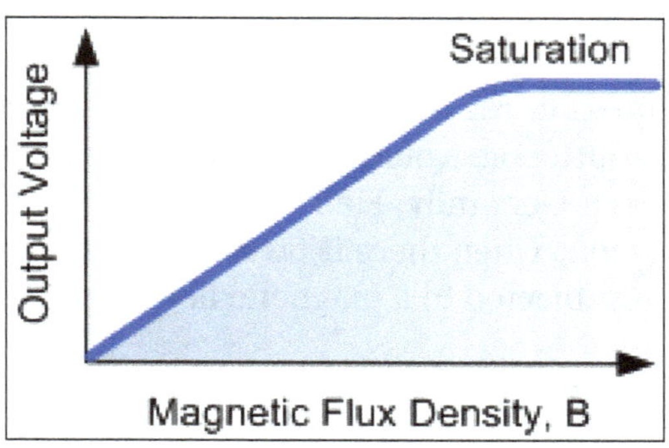

Linear or analogue sensors give a continuous voltage output that increases with a strong magnetic field and decreases with a weak magnetic field. In linear output Hall effect sensors, as the strength of the magnetic field increases the output signal from the amplifier will also increase until it begins to saturate by the limits imposed on it by the power supply. Any additional increase in the magnetic field will have no effect on the output but drive it more into saturation.

Digital output sensors on the other hand have a Schmitt-trigger with built in hysteresis connected to the op-amp. When the magnetic flux passing through the Hall sensor exceeds a pre-set value the output from the device switches quickly between its "off" condition to an "on" condition without any type of contact bounce. This built-in hysteresis eliminates any oscillation of the output signal as the sensor moves in and out of the magnetic field. Then digital output sensors have just two states, "on" and "off".

There are two basic types of digital Hall effect sensor, Bipolar and Unipolar. Bipolar sensors require a positive magnetic field (south pole) to operate them and a negative field (north pole) to release them while unipolar sensors require only a single magnetic south pole to both operate and release them as they move in and out of the magnetic field.

Most Hall effect devices can not directly switch large electrical loads as their output drive capabilities are very small around 10 to 20mA. For large current loads an open-collector (current sinking) NPN Transistor is added to the output.

This transistor operates in its saturated region as a NPN sink switch which shorts the output terminal to ground whenever the applied flux density is higher than that of the "on" pre-set point.

The output switching transistor can be either an open emitter transistor, open collector transistor configuration or both providing a push-pull output type configuration that can sink enough current to directly drive many loads, including relays, motors, LEDs, and lamps.

Hall Effect Applications

Hall effect sensors are activated by a magnetic field and in many applications the device can be operated by a single permanent magnet attached to a moving shaft or device. There are many different types of magnet movements, such as "Head-on", "Sideways", "Push-pull" or "Push-push" etc sensing movements. Which every type of configuration is used, to ensure maximum sensitivity the magnetic lines of flux must always be perpendicular to the sensing area of the device and must be of the correct polarity.

Also to ensure linearity, high field strength magnets are required that produce a large change in field strength for the required movement. There are several possible paths of motion for detecting a magnetic field, and below are two of the more common sensing configurations using a single magnet: Head-on Detection and Sideways Detection.

Head-on Detection

As its name implies, "head-on detection" requires that the magnetic field is perpendicular to the hall effect sensing device and that for detection, it approaches the sensor straight on towards the active face. A sort of "head-on" approach.

This head-on approach generates an output signal, VH which in the linear devices represents the strength of the magnetic field, the magnetic flux density, as a function of distance away from the hall effect sensor. The nearer and therefore the stronger the magnetic field, the greater the output voltage and vice versa.

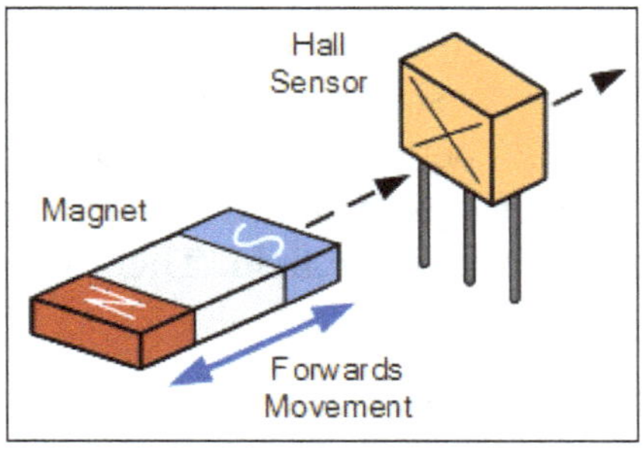

Linear devices can also differentiate between positive and negative magnetic fields. Non-linear devices can be made to trigger the output "ON" at a pre-set air gap distance away from the magnet for indicating positional detection.

Sideways Detection

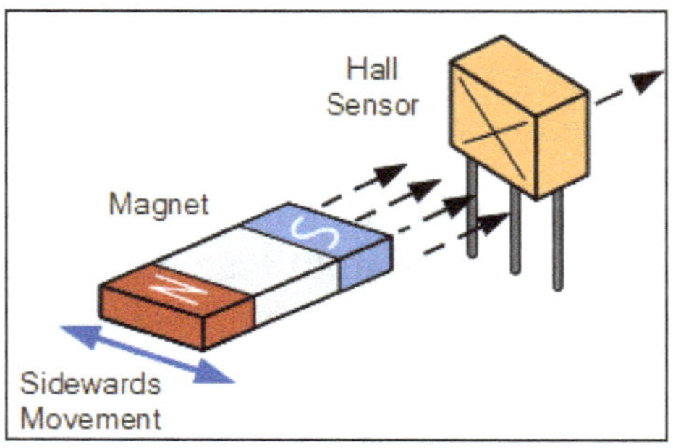

The second sensing configuration is "sideways detection". This requires moving the magnet across the face of the Hall effect element in a sideways motion.

Sideways or slide-by detection is useful for detecting the presence of a magnetic field as it moves across the face of the Hall element within a fixed air gap distance for example, counting rotational magnets or the speed of rotation of motors.

Depending upon the position of the magnetic field as it passes by the zero field centre line of the sensor, a linear output voltage representing both a positive and a negative output can be produced. This allows for directional movement detection which can be vertical as well as horizontal.

There are many different applications for Hall Effect Sensors especially as proximity sensors. They can be used instead of optical and light sensors were the environmental conditions consist of water, vibration, dirt or oil such as in automotive applications. Hall effect devices can also be used for current sensing.

When a current passes through a conductor, a circular electromagnetic field is produced around it. By placing the Hall sensor next to the conductor, electrical currents from a few milliamps into thousands of amperes can be measured from the generated magnetic field without the need of large or expensive transformers and coils.

As well as detecting the presence or absence of magnets and magnetic fields, Hall effect sensors can also be used to detect ferromagnetic materials such as iron and steel by placing a small permanent "biasing" magnet behind the active area of the device. The sensor now sits in a permanent and static magnetic field, and any change or disturbance to this magnetic field by the introduction of a ferrous material will be detected with sensitivities as low as mV/G possible.

There are many different ways to interface Hall effect sensors to electrical and electronic circuits depending upon the type of device, whether digital or linear. One very simple and easy to construct example is using a Light Emitting Diode as shown below.

Positional Detector

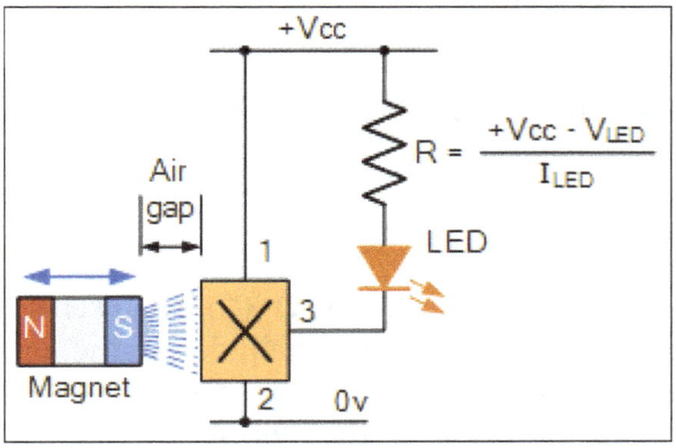

This head-on positional detector will be "off" when there is no magnetic field present, (0 gauss). When the permanent magnets south pole (positive gauss) is moved perpendicular towards the active area of the Hall effect sensor the device turns "on" and lights the LED. Once switched "on" the Hall effect sensor stays "on".

To turn the device and therefore the LED "off" the magnetic field must be reduced to below the release point for unipolar sensors or exposed to a magnetic north pole (negative gauss) for bipolar sensors. The LED can be replaced with a larger power transistor if the output of the Hall Effect Sensor is required to switch larger current loads.

Oxygen Sensor

An oxygen sensor (or lambda sensor, where lambda refers to air–fuel equivalence ratio, usually denoted by λ) is an electronic device that measures the proportion of oxygen (O_2) in the gas or liquid being analysed.

It was developed by Robert Bosch GmbH during the late 1960s under the supervision of Dr. Günter Bauman. The original sensing element is made with a thimble-shaped zirconia ceramic coated on both the exhaust and reference sides with a thin layer of platinum and comes in both heated and unheated forms. The planar-style sensor entered the market in 1990 and significantly reduced the mass of the ceramic sensing element, as well as incorporating the heater within the ceramic structure. This resulted in a sensor that started sooner and responded faster.

The most common application is to measure the exhaust-gas concentration of oxygen for internal combustion engines in automobiles and other vehicles in order to calculate and, if required, dynamically adjust the air-fuel ratio so that catalytic converters can work optimally, and also determine whether the converter is performing properly or not. Divers also use a similar device to measure the partial pressure of oxygen in their breathing gas.

Scientists use oxygen sensors to measure respiration or production of oxygen and use a different approach. Oxygen sensors are used in oxygen analyzers, which find extensive use in medical applications such as anesthesia monitors, respirators and oxygen concentrator so.

Oxygen sensors are also used in hypoxic air fire prevention systems to continuously monitor the oxygen concentration inside the protected volumes.

There are many different ways of measuring oxygen. These include technologies such as zirconia, electrochemical (also known as galvanic), infrared, ultrasonic, paramagnetic, and very recently, laser methods.

Automotive Applications

Automotive oxygen sensors, colloquially known as O_2 ("ō two") sensors, make modern electronic fuel injection and emission control possible. They help determine, in real time, whether the air–fuel ratio of a combustion engine is rich or lean. Since oxygen sensors are located in the exhaust stream, they do not directly measure the air or the fuel entering the engine, but when information from oxygen sensors is coupled with information from other sources, it can be used to indirectly determine the air–fuel ratio. Closed-loop feedback-controlled fuel injection varies the fuel injector output according to real-time sensor data rather than operating with a predetermined (open-loop) fuel map. In addition to enabling electronic fuel injection to work efficiently, this emissions control technique can reduce the amounts of both unburnt fuel and oxides of nitrogen entering the atmosphere. Unburnt fuel is pollution in the form of air-borne hydrocarbons, while oxides of nitrogen (NO_x gases) are a result of combustion chamber temperatures exceeding 1300 kelvins, due to excess air in the fuel mixture therefore contribute to smog and acid rain. Volvo was the first automobile manufacturer to employ this technology in the late 1970s, along with the three-way catalyst used in the catalytic converter.

The sensor does not actually measure oxygen concentration, but rather the difference between the amount of oxygen in the exhaust gas and the amount of oxygen in air. Rich mixture causes an oxygen demand. This demand causes a voltage to build up, due to transportation of oxygen ions through the sensor layer. Lean mixture causes low voltage, since there is an oxygen excess.

Modern spark-ignited combustion engines use oxygen sensors and catalytic converters in order to reduce exhaust emissions. Information on oxygen concentration is sent to the engine management computer or engine control unit (ECU), which adjusts the amount of fuel injected into the engine to compensate for excess air or excess fuel. The ECU attempts to maintain, on average, a certain air-fuel ratio by interpreting the information gained from the oxygen sensor. The primary goal is a compromise between power, fuel economy, and emissions, and in most cases is achieved by an air–fuel ratio close to stoichiometric. For spark-ignition engines (such as those that burn gasoline or LPG, as opposed to diesel), the three types of emissions modern systems are concerned with are: hydrocarbons (which are released when the fuel is not burnt completely, such as when misfiring or running rich), carbon monoxide (which is the result of running slightly rich) and NO_x (which dominate when the mixture is lean). Failure of these sensors, either through normal aging, the use of leaded fuels, or fuel contaminated with silicones or silicates, for example, can lead to damage of an automobile's catalytic converter and expensive repairs.

Tampering with or modifying the signal that the oxygen sensor sends to the engine computer can be detrimental to emissions control and can even damage the vehicle. When the engine is under low-load conditions (such as when accelerating very gently or maintaining a constant speed), it is operating in "closed-loop mode". This refers to a feedback loop between the ECU and the oxygen sensor(s) in which the ECU adjusts the quantity of fuel and expects to see a resulting change in the response of the oxygen sensor. This loop forces the engine to operate both slightly lean and slightly rich on successive loops, as it attempts to maintain a mostly stoichiometric ratio on average. If modifications cause the engine to run moderately lean, there will be a slight increase in fuel efficiency, sometimes at the expense of increased NO_x emissions, much higher exhaust gas temperatures, and sometimes a slight increase in power that can quickly turn into misfires and a drastic loss of power, as well as potential engine and catalytic-converter (due to the misfires) damage, at ultra-lean air–fuel ratios. If modifications cause the engine to run rich, then there will be a slight increase in power to a point (after which the engine starts flooding from too much unburned fuel), but at the cost of decreased fuel efficiency, and an increase in unburned hydrocarbons in the exhaust, which causes overheating of the catalytic converter. Prolonged operation at rich mixtures can cause catastrophic failure of the catalytic converter. The ECU also controls the spark engine timing along with the fuel-injector pulse width, so modifications that alter the engine to operate either too lean or too rich may result in inefficient fuel consumption whenever fuel is ignited too soon or too late in the combustion cycle.

When an internal combustion engine is under high load (e.g. wide open throttle), the output of the oxygen sensor is ignored, and the ECU automatically enriches the mixture to protect the engine, as misfires under load are much more likely to cause damage. This is referred to as an engine running in "open-loop mode". Any changes in the sensor output will be ignored in this state. In many cars (with the exception of some turbocharged models), inputs from the air flow meter are also ignored, as they might otherwise lower engine performance due to the mixture being too rich or too lean, and increase the risk of engine damage due to detonation if the mixture is too lean.

Function of a Lambda Probe

Lambda probes are used to reduce vehicle emissions by ensuring that engines burn their fuel efficiently and cleanly. Robert Bosch GmbH introduced the first automotive lambda probe in 1976, and it was first used by Volvo and Saab in that year. The sensors were introduced in the US from about 1979 and were required on all models of cars in many countries in Europe in 1993.

By measuring the proportion of oxygen in the remaining exhaust gas, and by knowing the volume and temperature of the air entering the cylinders amongst other things, an ECU can use look-up tables to determine the amount of fuel required to burn at the stoichiometric ratio (14.7:1 air:fuel by mass for gasoline) to ensure complete combustion.

Probe

The sensor element is a ceramic cylinder plated inside and outside with porous platinum electrodes; the whole assembly is protected by a metal gauze. It operates by measuring the difference in oxygen between the exhaust gas and the external air and generates a voltage or changes its resistance depending on the difference between the two.

The sensors only work effectively when heated to approximately 316 °C (600 °F), so most newer lambda probes have heating elements encased in the ceramic that bring the ceramic tip up to temperature quickly. Older probes, without heating elements, would eventually be heated by the exhaust, but there is a time lag between when the engine is started and when the components in the exhaust system come to a thermal equilibrium. The length of time required for the exhaust gases to bring the probe to temperature depends on the temperature of the ambient air and the geometry of the exhaust system. Without a heater, the process may take several minutes. There are pollution problems that are attributed to this slow start-up process, including a similar problem with the working temperature of a catalytic converter.

The probe typically has four wires attached to it: two for the lambda output, and two for the heater power, although some automakers use the metal case as ground for the sensor element signal, resulting in three wires. Earlier non-electrically-heated sensors had one or two wires.

Operation of the Probe

Zirconia Sensor

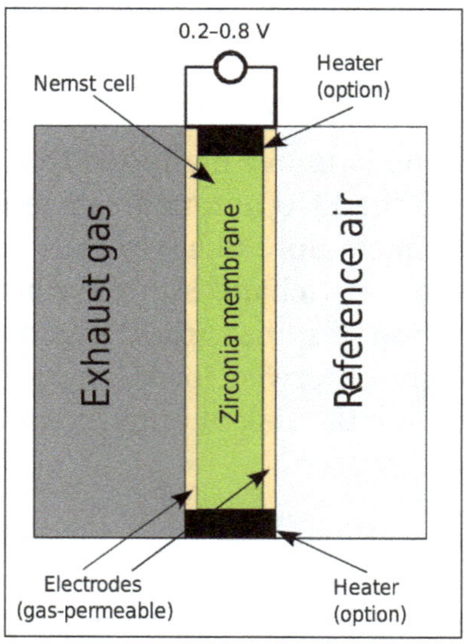

A planar zirconia sensor.

The zirconium dioxide, or zirconia, lambda sensor is based on a solid-state electrochemical fuel cell called the Nernst cell. Its two electrodes provide an output voltage corresponding to the quantity of oxygen in the exhaust relative to that in the atmosphere.

An output voltage of 0.2 V (200 mV) DC represents a "lean mixture" of fuel and oxygen, where the amount of oxygen entering the cylinder is sufficient to fully oxidize the carbon monoxide (CO), produced in burning the air and fuel, into carbon dioxide (CO_2). An output voltage of 0.8 V (800 mV) DC represents a "rich mixture", which is high in unburned fuel and low in remaining oxygen. The ideal setpoint is approximately 0.45 V (450 mV) DC. This is where the quantities of air and fuel are in the optimal ratio, which is ~0.5% lean of the stoichiometric point, such that the exhaust output contains minimal carbon monoxide.

The voltage produced by the sensor is nonlinear with respect to oxygen concentration. The sensor is most sensitive near the stoichiometric point (where $\lambda = 1$) and less sensitive when either very lean or very rich.

The ECU is a control system that uses feedback from the sensor to adjust the fuel/air mixture. As in all control systems, the time constant of the sensor is important; the ability of the ECU to control the fuel–air ratio depends upon the response time of the sensor. An aging or fouled sensor tends to have a slower response time, which can degrade system performance. The shorter the time period, the higher the so-called "cross count" and the more responsive the system.

The sensor has a rugged stainless-steel construction internally and externally. Due to this the sensor has a high resistance to corrosion, allowing it to be used effectively in aggressive environments with high temperature/pressure.

The zirconia sensor is of the "narrow-band" type, referring to the narrow range of fuel/air ratios to which it responds.

Wideband Zirconia Sensor

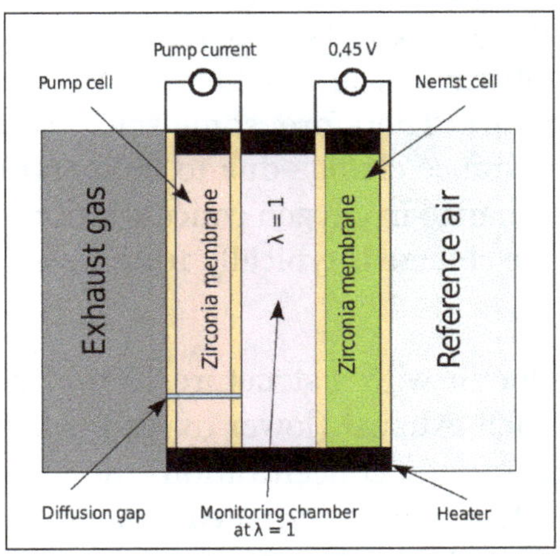

A planar wideband zirconia sensor.

A variation on the zirconia sensor, called the "wideband" sensor, was introduced by NTK in 1992 and has been widely used for car engine management systems in order to meet the ever-increasing demands for better fuel economy, lower emissions and better engine performance at the same time. It is based on a planar zirconia element, but also incorporates an electrochemical gas pump. An electronic circuit containing a feedback loop controls the gas-pump current to keep the output of the electrochemical cell constant, so that the pump current directly indicates the oxygen content of the exhaust gas. This sensor eliminates the lean–rich cycling inherent in narrow-band sensors, allowing the control unit to adjust the fuel delivery and ignition timing of the engine much more rapidly. In the automotive industry this sensor is also called a *UEGO* (universal exhaust-gas oxygen) sensor.

UEGO sensors are also commonly used in aftermarket dyno tuning and high-performance driver air–fuel display equipment. The wideband zirconia sensor is used in stratified fuel injection systems and can now also be used in diesel engines to satisfy the upcoming EURO and ULEV emission limits.

Wideband sensors have three elements:

- Ion oxygen pump,
- Narrowband zirconia sensor,
- Heating element.

The wiring diagram for the wideband sensor typically has five wires:

- Resistive heating element,
- Sensor,
- Pump,
- Calibration resistor,
- Common.

Titania Sensor

A less common type of narrow-band lambda sensor has a ceramic element made of titania (titanium dioxide). This type does not generate its own voltage, but changes its electrical resistance in response to the oxygen concentration. The resistance of the titania is a function of the oxygen partial pressure and the temperature. Therefore, some sensors are used with a gas-temperature sensor to compensate for the resistance change due to temperature. The resistance value at any temperature is about 1/1000 the change in oxygen concentration. Luckily, at $\lambda = 1$, there is a large change of oxygen, so the resistance change is typically 1000 times between rich and lean, depending on the temperature.

As titania is an N-type semiconductor with a structure TiO_{2-x}, the x defects in the crystal lattice conduct the charge. So, for fuel-rich exhaust (lower oxygen concentration) the resistance is low, and for fuel-lean exhaust (higher oxygen concentration) the resistance is high. The control unit feeds the sensor with a small electric current and measures the resulting voltage drop across the sensor, which varies from nearly 0 volts to about 5 volts. Like the zirconia sensor, this type is nonlinear, such that it is sometimes simplistically described as a binary indicator, reading either "rich" or "lean". Titania sensors are more expensive than zirconia sensors, but they also respond faster.

In automotive applications the titania sensor, unlike the zirconia sensor, does not require a reference sample of atmospheric air to operate properly. This makes the sensor assembly easier to design against water contamination. While most automotive sensors are submersible, zirconia-based sensors require a very small supply of reference air from the atmosphere. In theory, the sensor wire harness and connector are sealed. Air that leaches through the wire harness to the sensor is assumed to come from an open point in the harness – usually the ECU, which is housed in an enclosed space like the trunk or vehicle interior.

Location of the Probe in a System

The probe is typically screwed into a threaded hole in the exhaust system, located after the branch manifold of the exhaust system combines and before the catalytic converter. New vehicles are required to have a sensor before and after the exhaust catalyst to meet U.S. regulations requiring that all emissions components be monitored for failure. Pre-and post-catalyst signals are monitored to determine catalyst efficiency, and if the converter is not performing as expected, an alert gets reported to the user through on-board diagnostics systems by, for example, lighting up an indicator in the vehicle's dashboard. Additionally, some catalyst systems require brief cycles of lean (oxygen-containing) gas to load the catalyst and promote additional oxidation reduction of undesirable exhaust components.

Sensor Surveillance

The air–fuel ratio and naturally, the status of the sensor, can be monitored by means of using an air–fuel ratio meter that displays the output voltage of the sensor.

Sensor Failures

Normally, the lifetime of an unheated sensor is about 30,000 to 50,000 miles (50,000 to 80,000 km). Heated sensor lifetime is typically 100,000 miles (160,000 km). Failure of an unheated sensor is usually caused by the buildup of soot on the ceramic element, which lengthens its response time and may cause total loss of ability to sense oxygen. For heated sensors, normal deposits are burned off during operation, and failure occurs due to catalyst depletion. The probe then tends to report lean mixture, the ECU enriches the mixture, the exhaust gets rich with carbon monoxide and hydrocarbons, and the fuel economy worsens.

Leaded gasoline contaminates the oxygen sensors and catalytic converters. Most oxygen sensors are rated for some service life in the presence of leaded gasoline, but sensor life will be shortened to as little as 15,000 miles (24,000 km), depending on the lead concentration. Lead-damaged sensors typically have their tips discolored light rusty.

Another common cause of premature failure of lambda probes is contamination of fuel with silicones (used in some sealings and greases) or silicates (used as corrosion inhibitors in some antifreezes). In this case, the deposits on the sensor are colored between shiny white and grainy light gray.

Leaks of oil into the engine may cover the probe tip with an oily black deposit, with associated loss of response.

An overly rich mixture causes buildup of black powdery deposit on the probe. This may be caused by failure of the probe itself, or by a problem elsewhere in the fuel-rationing system.

Applying an external voltage to the zirconia sensors, e.g. by checking them with some types of ohmmeter, may damage them.

Some sensors have an air inlet to the sensor in the lead, so contamination from the lead caused by water or oil leaks can be sucked into the sensor and cause failure.

Symptoms of a failing oxygen sensor includes:

- Sensor light on dash indicates problem,
- Increased tailpipe emissions,
- Increased fuel consumption,
- Hesitation on acceleration,
- Stalling,
- Rough idling.

Diving Applications

An oxygen analyser for breathing gas mixtures for diving.

The type of oxygen sensor used in most underwater diving applications is the electro-galvanic oxygen sensor, a type of fuel cell, which is sometimes called an oxygen analyser or ppO_2 meter. They are used to measure the oxygen concentration of breathing gas mixes such as nitrox and trimix. They are also used within the oxygen control mechanisms of closed-circuit rebreathers to keep the partial pressure of oxygen within safe limits. and to monitor the oxygen content of the breathing gas in saturation diving systems and of surface supplied mixed gas. This type of sensor operates by measuring the voltage generated by a small electro-galvanic fuel cell.

Scientific Applications

In soil respiration studies oxygen sensors can be used in conjunction with carbon dioxide sensors to help improve the characterization of soil respiration. Typically, soil oxygen sensors use a galvanic cell to produce a current flow that is proportional to the oxygen concentration being measured. These sensors are buried at various depths to monitor oxygen depletion over time, which is then used to predict soil respiration rates. Generally, these soil sensors are equipped with a built-in heater to prevent condensation from forming on the permeable membrane, as relative humidity can reach 100% in soil.

In marine biology or limnology, oxygen measurements are usually done in order to measure respiration of a community or an organism, but have also been used to measure primary production of algae. The traditional way of measuring oxygen concentration in a water sample has been to use wet chemistry techniques e.g. the Winkler titration method. There are however commercially available oxygen sensors that measure the oxygen concentration in liquids with great accuracy. There are two types of oxygen sensors available: electrodes (electrochemical sensors) and optodes (optical sensors).

Electrodes

A dissolved oxygen meter for laboratory use.

The Clark-type electrode is the most used oxygen sensor for measuring oxygen dissolved in a liquid. The basic principle is that there is a cathode and an anode submersed in an electrolyte. Oxygen enters the sensor through a permeable membrane by diffusion and is reduced at the cathode, creating a measurable electric current.

There is a linear relationship between the oxygen concentration and the electric current. With a two-point calibration (0% and 100% air saturation), it is possible to measure oxygen in the sample.

One drawback to this approach is that oxygen is consumed during the measurement with a rate equal to the diffusion in the sensor. This means that the sensor must be stirred in order to get the correct measurement and avoid stagnant water. With an increasing sensor size, the oxygen consumption increases and so does the stirring sensitivity. In large sensors there tend to also be a drift in the signal over time due to consumption of the electrolyte. However, Clark-type sensors can be made very small with a tip size of 10 µm. The oxygen consumption of such a microsensor is so small that it is practically insensitive to stirring and can be used in stagnant media such as sediments or inside plant tissue.

Optodes

An oxygen optode is a sensor based on optical measurement of the oxygen concentration. A chemical film is glued to the tip of an optical cable, and the fluorescence properties of this film depend on the oxygen concentration. Fluorescence is at a maximum when there is no oxygen present. When an O_2 molecule comes along, it collides with the film, and this quenches the photoluminescence. In a given oxygen concentration there will be a specific number of O_2 molecules colliding with the film at any given time, and the fluorescence properties will be stable.

The signal (fluorescence) to oxygen ratio is not linear, and an optode is most sensitive at low oxygen concentration. That is, the sensitivity decreases as oxygen concentration increases, following the Stern–Volmer relationship. The optode sensors can, however, work in the whole region 0% to 100% oxygen saturation in water, and the calibration is done the same way as with the Clark-type sensor. No oxygen is consumed, and hence the sensor is insensitive to stirring, but the signal will stabilize more quickly if the sensor is stirred after being put in the sample. These type of electrode sensors can be used for in situ and real-time monitoring of oxygen production in water-splitting reactions. The platinized electrodes can accomplish the real-time monitoring of hydrogen production in water-splitting device.

Wheel Speed Sensor

A wheel speed sensor or vehicle speed sensor (VSS) is a type of tachometer. It is a sender device used for reading the speed of a vehicle's wheel rotation. It usually consists of a toothed ring and pickup.

Automotive Wheel Speed Sensors

Purpose

The wheel speed sensor was initially used to replace the mechanical linkage from the road wheels to the speedometer, eliminating cable breakage and simplifying the gauge construction (elimination all moving parts except for the needle/spring assembly). With the advent of automated driving aid, such as electronic ABS, the sensor also provided wheel speed data to the controllers to assist the operator in maintaining control of the vehicle. The vehicle Speed sensor is also used for the proper shifting up of gears for the vehicle maintenance.

Construction

The most common wheel speed sensor system consists of a ferromagnetic toothed reluctor ring (tone wheel) and a sensor (which can be passive or active).

The tone wheel is typically made of steel and may be an open-air design, or sealed (as in the case of unitized bearing assemblies). The number of teeth is chosen as a trade-off between low-speed sensing/accuracy and high-speed sensing/cost. Greater numbers of teeth will require more machining operations and (in the case of passive sensors) produce a higher frequency output signal which may not be as easily interpreted at the receiving end, but give a better resolution and higher signal update rate. In more advanced systems, the teeth can be asymmetrically shaped to allow the sensor to distinguish between forward and reverse rotation of the wheel.

A passive sensor typically consists of a ferromagnetic rod which is oriented to project radially from the tone wheel with a permanent magnet at the opposite end. The rod is wound with fine wire which experiences an induced alternating voltage as the tone wheel rotates, as the teeth interfere with the magnetic field. Passive sensors output a sinusoidal signal which grows in magnitude and frequency with wheel speed.

A variation of the passive sensor does not have a magnet backing it, but rather a tone wheel which consists of alternating magnetic poles produce the alternating voltage. The output of this sensor tends to resemble a square wave, rather than a sinusoid, but still increases in magnitude as wheels speed increases.

An active sensor is a passive sensor with signal conditioning circuitry built into the device. This signal conditioning may be amplifying the signal's magnitude; changing the signal's form to PWM, square wave, or others; or encoding the value into a communication protocol (such as CAN) before transmission.

Variations

The vehicle speed sensor (VSS) may be, but is not always, a true wheel speed sensor. For example, in the Ford AOD transmission, the VSS is mounted to the tailshaft extension housing and is a self-contained tone ring and sensor. Though this does not give wheel speed (as each wheel in an axle with a differential is able to turn at differing speeds, and neither is solely dependent on the driveshaft for its final speed), under typical driving conditions this is close enough to provide the speedometer signal, and was used for the rear wheel ABS systems on 1987 and newer Ford F-Series, the first pickups with ABS.

Special Purpose Speed Sensors

Road Vehicles

Wheel speed sensors are in anti-lock braking systems in conjunction with the Electronic Stability Program(Control) system.

Rotary Speed Sensors for Rail Vehicles

Many of the subsystems in a rail vehicle, such as a locomotive or multiple unit, depend on a reliable and precise rotary speed signal, in some cases as a measure of the speed or changes in the speed. This applies in particular to traction control, but also to wheel slide protection, registration, train control, door control and so on. These tasks are performed by a number of rotary speed sensors that may be found in various parts of the vehicle.

Speed sensor failures are frequent, and are mainly due to the extremely harsh operating conditions encountered in rail vehicles. The relevant standards specify detailed test criteria, but in practical operation the conditions encountered are often even more extreme (such as shock/vibration and especially electromagnetic compatibility (EMC)).

Rotary Speed Sensors for Motors

Although rail vehicles occasionally do use drives without sensors, most need a rotary speed sensor for their regulator system. The most common type is a two-channel sensor that scans a toothed wheel on the motor shaft or gearbox which may be dedicated to this purpose or may be already present in the drive system.

Modern Hall effect sensors of this type make use of the principle of magnetic field modulation and are suitable for ferromagnetic target wheels with a module between m =1 and m = 3.5 (D.P.=25 to

D.P.=7). The form of the teeth is of secondary importance; target wheels with involute or rectangular toothing can be scanned. Depending on the diameter and teeth of the wheel it is possible to get between 60 and 300 pulses per revolution, which is sufficient for drives of lower and medium traction performance.

This type of sensor normally consists of two hall effect sensors, a rare earth magnet and appropriate evaluation electronics. The field of the magnet is modulated by the passing target teeth. This modulation is registered by the Hall sensors, converted by a comparator stage to a square wave signal and amplified in a driver stage.

Unfortunately, the Hall effect varies greatly with temperature. The sensors' sensitivity and also the signal offset therefore depend not only on the air gap but also on the temperature. This also very much reduces the maximum permissible air gap between the sensor and the target wheel. At room temperature an air gap of 2 to 3 mm can be tolerated without difficulty for a typical target wheel of module m = 2, but in the required temperature range of from −40 °C to 120 °C the maximum gap for effective signal registration drops to 1.3 mm. Smaller pitch target wheels with module m = 1 are often used to get a higher time resolution or to make the construction more compact. In this case the maximum possible air gap is only 0.5 to 0.8 mm.

For the design engineer, the visible air gap that the sensor ends up with is primarily the result of the specific machine design, but is subject to whatever constraints are needed to register the rotary speed. If this means that the possible air gap has to lie within a very small range, then this will also restrict the mechanical tolerances of the motor housing and target wheels to prevent signal dropouts during operation. This means that in practice there may be problems, particularly with smaller pitched target wheels of module m = 1 and disadvantageous combinations of tolerances and extreme temperatures. From the point of view of the motor manufacturer, and even more so the operator, it is therefore better to look for speed sensors with a wider range of air gap.

The primary signal from a Hall sensor loses amplitude sharply as the air gap increases. For Hall sensor manufacturers this means that they need to provide maximum possible compensation for the Hall signal's physically induced offset drift. The conventional way of doing this is to measure the temperature at the sensor and use this information to compensate the offset, but this fails for two reasons: firstly because the drift does not vary linearly with the temperature, and secondly because not even the sign of the drift is the same for all sensors.

Some sensors now offer an integrated signal processor that attempts to correct the offset and amplitude of the Hall sensor signals. This correction enables a larger maximum permissible air gap at the speed sensor. On a module m = 1 target wheel these new sensors can tolerate an air gap of 1.4 mm, which is wider than that for conventional speed sensors on module m = 2 target wheels. On a module m = 2 target wheel the new speed sensors can tolerate gap of as much as 2.2 mm. It has also been possible to markedly increase the signal quality. Both the duty cycle and the phase displacement between the two channels is at least three times as stable in the face of fluctuating air gap and temperature drift. In addition, in spite of the complex electronics it has also been possible to increase the mean time between failures for the new speed sensors by a factor of three to four. So they not only provide more precise signals, their signal availability is also significantly better.

An alternative to Hall effect sensors with gears are sensors or encoders which use [magnetoresistance]. Because the target wheel is an active, multipole magnet, air gaps can be even larger, up to

4.0 mm. Because magnetoresistive sensors are angle-sensitive and amplitude-insensitive, signal quality is increased over Hall sensors in fluctuating gap applications. Also the signal quality is much higher, enabling [interpolation] within the sensor/encoder or by an external circuit.

Motor Encoders with Integrated Bearings

There is a limit on the number of pulses achievable by Hall sensors without integrated bearings: with a 300 mm diameter target wheel it is normally not possible to get beyond 300 pulses per revolution. But many locomotives and electric multiple units (EMUs) need higher numbers of pulses for proper operation of the traction converter, for instance when there are tight constraints on the traction regulator at low speeds.

Such Hall effect sensor applications may benefit from built-in bearings, which can tolerate an air gap many orders of magnitude smaller because of the greatly reduced play on the actual sensor as opposed to that of the motor bearing. This makes it possible to choose a much smaller pitch for the measuring scale, right down to module m = 0.22. Likewise, the magnetoresistive sensors offer even higher resolution and accuracy than Hall sensors when implemented in motor encoders with integrated bearings.

For even greater signal accuracy a precision encoder can be used. The functional principles of the two encoders are similar: a multichannel magneto-resistive sensor scans a target wheel with 256 teeth, generating sine and cosine signals. Arctangent interpolation is used to generate rectangular pulses from the sine/cosine signal periods. The precision encoder also possesses amplitude and offset correction functions. This makes it possible to further improve the signal quality, which greatly improves traction regulation.

Speed Sensors on the Wheelset

Bearingless Wheelset Speed Sensors

Bearingless speed sensors may be found in almost every wheelset of a rail vehicle. They are principally used for wheel slide protection and usually supplied by the manufacturer of the wheel slide protection system. These sensors require a sufficiently small air gap and need to be particularly reliable. One special feature of rotary speed sensors that are used for wheel slide protection is their integrated monitoring functions. Two-wire sensors with a current output of 7 mA/14 mA are used to detect broken cables. Other designs provide for an output voltage of around 7 V as soon as the signal frequency drops below 1 Hz. Another method used is to detect a 50 MHz output signal from the sensor when the power supply is periodically modulated at 50 MHz. It is also common for two-channel sensors to have electrically isolated channels.

Occasionally it is necessary to take off the wheel slide protection signal at the traction motor, and the output frequency is then often too high for the wheel slide protection electronics. For this application a speed sensor with an integrated frequency divider or encoder can be utilized.

Wheelset Pulse Generator with Integrated Bearing

A rail vehicle, particularly a locomotive, possesses numerous subsystems that require separate, electrically isolated speed signals. There usually are neither enough mounting places nor is there sufficient space where separate pulse generators could be installed. Multi-channel pulse generators that are flange-mounted onto the bearing shells or covers of wheelsets offer a solution. Using a number of bearingless speed sensors would also involve additional cables, which should preferably be avoided for outdoor equipment because they are so susceptible to damage, for instance from flying track ballast.

Optical Sensor

From one to four channels can be implemented, each channel having a photosensor that scans one of at most two signal tracks on a slotted disk. Experience shows that the possible number of channels achievable by this technique is still not enough. A number of subsystems therefore have to make do with looped-through signals from the wheel slide protection electronics and are therefore forced to accept, for instance, the available number of pulses, although a separate speed signal might well have some advantages.

The use of optical sensors is widespread in industry. Unfortunately they do have two fundamental weaknesses that have always made it very difficult to get them to function reliably over a number of years, namely – the optical components are extremely susceptible to dirt, and – the light source ages too quickly.

Even traces of dirt greatly reduce the amount of light that passes through the lens and can cause signal dropout. These encoders are therefore required to be very well sealed. Further problems are encountered when the pulse generators are used in environments in which the dew point is passed: the lenses fog and the signal is frequently interrupted.

The light sources used are light-emitting diodes (LEDs). But LEDs are always subject to aging, which over a few years leads to a noticeably reduced beam. Attempts are made to compensate for this by using special regulators that gradually increase the current through the LED, but unfortunately this further accelerates the aging process.

Magnetic Sensor

The principle used in scanning a ferromagnetic measuring scale magnetically does not exhibit these deficiencies. During many years' experience of using magnetic encoders there have been occasions when a seal has failed and a pulse generator has been found to be completely covered in a thick layer of brake dust and other dirt, but such pulse generators still functioned perfectly.

Historically, magnetic sensor systems cost more than optical systems, but this difference is narrowing rapidly. Magnetic Hall and magnetoresistive sensor systems can be imbedded in plastic or potting material, which increases mechanical reliability and eliminates damage from water and grease.

Wheel speed sensors can also include hysteresis. This suppresses any extraneous pulses while the vehicle is at a standstill.

Pulse generators constructed in accordance with this principle have been successfully field tested by several rail operators since the beginning of 2005. The type test specified in EN 50155 has also been successfully completed, so that these pulse generators can now be delivered.

Wheelset Pulse Generators for Inside-Journal Bogies

Inside-journal bogies make particular demands on the pulse generator designer because they have no bearing cover on the end to serve as the basis from which the rotation of the wheelset shaft could be registered. In this case the pulse generator has to be mounted on a shaft stub attached to the wheelset and fitted with a torque converter connected to the bogie frame to prevent it from rotating.

The extreme vibration in this location leads to a considerable load on the pulse generator bearing, which, with this method of installation has to carry not only the relatively small mass of the pulse generator shaft but that of the entire pulse generator. When we consider that bearing life reduces with at least the third power of the load we can see that a reliable and durable pulse generator for such a situation cannot merely be adapted from the more common standard pulse generator for outside-journal bogies merely by fitting and intermediate flange or similar construction. It really is necessary to have a pulse generator with a modified design adapted to the requirements of such a location.

Speed Sensors for Non-magnetic Target Wheels

Some transport companies are faced with a special problem: the circulating air that keeps the motors cool carries swarf abraded from the wheels and rails. This collects on the heads of magnetic sensors. There are also increasingly motors in which sensors have to scan aluminium target wheels, for instance because the impellers are made of an aluminium alloy and the manufacturer does not wish to have to shrink on a separate ferromagnetic gear rim.

For these applications there are speed sensors available that do not require a target magnet. A number of transmitting and receiving coils are used to generate an alternating electric field with a frequency of the order of 1 MHz and the modulation of the coupling between senders and receivers is then evaluated. This sensor is installation and signal compatible to the magnetic sensors; for most common target wheel modules the units can simply be replaced without any other measures being necessary.

Speed Sensors with Interpolation

Customers often want a higher number of pulses per revolution than can be achieved in the space available and with the smallest module m = 1. To achieve this goal, sensors are available which offer interpolation. These offer output of 2-64X the original number of gear teeth or magnetic poles on the target wheel. Accuracy is dependent on the quality of sensor input: Hall sensors are lower cost, but lower accuracy, magnetoresistive sensors are higher cost, but higher accuracy.

Torque Sensor

A torque sensor, torque transducer or torque meter is a device for measuring and recording the torque on a rotating system, such as an engine, crankshaft, gearbox, transmission, rotor, a bicycle crank or cap torque tester. Static torque is relatively easy to measure. Dynamic torque, on the other hand, is not easy to measure, since it generally requires transfer of some effect (electric, hydraulic or magnetic) from the shaft being measured to a static system.

One way to achieve this is to condition the shaft or a member attached to the shaft with a series of permanent magnetic domains. The magnetic characteristics of these domains will vary according to the applied torque, and thus can be measured using non-contact sensors. Such magnetoelastic torque sensors are generally used for in-vehicle applications on racecars, automobiles, aircraft, and hovercraft.

Commonly, torque sensors or torque transducers use strain gauges applied to a rotating shaft or axle. With this method, a means to power the strain gauge bridge is necessary, as well as a means to receive the signal from the rotating shaft. This can be accomplished using slip rings, wireless telemetry, or rotary transformers. Newer types of torque transducers add conditioning electronics and an A/D converter to the rotating shaft. Stator electronics then read the digital signals and convert those signals to a high-level analog output signal, such as +/-10VDC.

A more recent development is the use of SAW devices attached to the shaft and remotely interrogated. The strain on these tiny devices as the shaft flexes can be read remotely and output without the need for attached electronics on the shaft. The probable first use in volume will be in the automotive field as, of May 2009, Schott announced it has a SAW sensor package viable for in vehicle uses.

Another way to measure torque is by way of twist angle measurement or phase shift measurement, whereby the angle of twist resulting from applied torque is measured by using two angular position sensors and measuring the phase angle between them. This technique is used in the Allison T56 turboprop engine.

Finally, if the mechanical system involves a right angle gearbox, then the axial reaction force experienced by the inputting shaft/pinion can be related to the torque experienced by the output shaft(s). The axial input stress must first be calibrated against the output torque. The input stress can be easily measured via strain gauge measurement of the input pinion bearing housing. The output torque is easily measured using a static torque meter.

Throttle Position Sensor

A throttle position sensor (TPS) is a sensor used to monitor the air intake of an engine. The sensor is usually located on the butterfly spindle/shaft so that it can directly monitor the position of the throttle. More advanced forms of the sensor are also used, for example an extra *closed throttle position sensor* (CTPS) may be employed to indicate that the throttle is completely closed. Some engine control units (ECUs) also control the throttle position electronic throttle control (ETC) or "drive by wire" systems and if that is done the position sensor is used in a feedback loop to enable that control.

Related to the TPS are accelerator pedal sensors, which often include a wide open throttle (WOT) sensor. The accelerator pedal sensors are used in electronic throttle control or "drive by wire" systems, and the most common use of a wide open throttle sensor is for the kick-down function on automatic transmissions.

Modern day sensors are non contact type. These modern non contact TPS include Hall effect sensors, inductive sensors, magnetoresistive and others. In the potentiometric type sensors, a multi-finger metal brush/rake is in contact with a resistive strip, while the butterfly valve is turned from the lower mechanical stop (minimum air position) to WOT, there is a change in the resistance and this change in resistance is given as the input to the ECU.

Non contact type TPS work on the principle of Hall effect or inductive sensors, or magnetoresistive technologies, wherein generally the magnet or inductive loop is the dynamic part which is mounted on the butterfly valve throttle spindle/shaft gear and the sensor & signal processing circuit board is mounted within the ETC gear box cover and is stationary. When the magnet/inductive loop mounted on the spindle which is rotated from the lower mechanical stop to WOT, there is a change in the magnetic field for the sensor. The change in the magnetic field is sensed by the sensor and the voltage generated is given as the input to the ECU. Normally a two pole rare-earth magnet is used for the TPS due to their high Curie temperatures required in the under-hood vehicle environment. The magnet may be of diametrical type, ring type, rectangular or segment type. The magnet is defined to have a certain magnetic field that does not vary significantly with time or temperature.

Soil Moisture Sensor

Soil moisture sensors measure the volumetric water content in soil. Since the direct gravimetric measurement of free soil moisture requires removing, drying, and weighting of a sample, soil moisture sensors measure the volumetric water content indirectly by using some other property of the soil, such as electrical resistance, dielectric constant, or interaction with neutrons, as a proxy for the moisture content.

The relation between the measured property and soil moisture must be calibrated and may vary depending on environmental factors such as soil type, temperature, or electric conductivity. Reflected microwave radiation is affected by the soil moisture and is used for remote sensing in hydrology and agriculture. Portable probe instruments can be used by farmers or gardeners.

Soil moisture sensors typically refer to sensors that estimate volumetric water content. Another class of sensors measure another property of moisture in soils called water potential; these sensors are usually referred to as soil water potential sensors and include tensiometers and gypsum blocks.

A simple soil moisture sensor for gardeners.

Technology

Technologies commonly used to indirectly measure volumetric water content (soil moisture) include:

- Frequency Domain Reflectometry (FDR): The dielectric constant of a certain volume element around the sensor is obtained by measuring the operating frequency of an oscillating circuit.
- Time Domain Transmission (TDT) and Time Domain Reflectometry (TDR): The dielectric constant of a certain volume element around the sensor is obtained by measuring the speed of propagation along a buried transmission line.
- Neutron moisture gauges: The moderator properties of water for neutrons are utilized to estimate soil moisture content between a source and detector probe.
- Soil resistivity: Measuring how strongly the soil resists the flow of electricity between two electrodes can be used to determine the soil moisture content.
- Galvanic cell: The amount of water present can be determined based on the voltage the soil produces because water acts as an electrolyte and produces electricity. The technology behind this concept is the galvanic cell.

Application

Agriculture

Measuring soil moisture is important for agricultural applications to help farmers manage their irrigation systems more efficiently. Knowing the exact soil moisture conditions on their fields, not only are farmers able to generally use less water to grow a crop, they are also able to increase yields and the quality of the crop by improved management of soil moisture during critical plant growth stages.

Landscape Irrigation

In urban and suburban areas, landscapes and residential lawns are using soil moisture sensors to interface with an irrigation controller. Connecting a soil moisture sensor to a simple irrigation clock will convert it into a "smart" irrigation controller that prevents irrigation cycles when the soil is already wet, e.g. following a recent rainfall event.

Golf courses are using soil moisture sensors to increase the efficiency of their irrigation systems to prevent over-watering and leaching of fertilizers and other chemicals into the ground.

Soil moisture sensors are used in numerous research applications, e.g. in agricultural science and horticulture including irrigation planning, climate research, or environmental science including solute transport studies and as auxiliary sensors for soil respiration measurements.

Simple Sensors for Gardeners

Relatively cheap and simple devices that do not require a power source are available for checking whether plants have sufficient moisture to thrive. After inserting a probe into the soil for approximately 60 seconds, a meter indicates if the soil is too dry, moist or wet for plants.

Active Pixel Sensor

An active-pixel sensor (APS) is an image sensor where each picture element ("pixel") has a photodetector and an active amplifier. There are many types of integrated circuit active pixel sensors including the complementary metal–oxide–semiconductor (CMOS) APS used most commonly in cell phone cameras, web cameras, most digital pocket cameras since 2010, in most digital single-lens reflex cameras (DSLRs) and Mirrorless interchangeable-lens cameras (MILCs). Such an image sensor is produced using CMOS technology (and is hence also known as a CMOS sensor), and has emerged as an alternative to charge-coupled device (CCD) image sensors.

CMOS image sensor.

The term 'active pixel sensor' is also used to refer to the individual pixel sensor itself, as opposed to the image sensor; in that case the image sensor is sometimes called an active pixel sensor imager, or active-pixel image sensor.

Comparison to CCDs

APS pixels solve the speed and scalability issues of the passive-pixel sensor. They generally consume less power than CCDs, have less image lag, and require less specialized manufacturing facilities. Unlike CCDs, APS sensors can combine the image sensor function and image processing functions within the same integrated circuit. APS sensors have found markets in many consumer applications, especially camera phones. They have also been used in other fields including digital radiography, military ultra high speed image acquisition, security cameras, and optical mice. Manufacturers include Aptina Imaging (independent spinout from Micron Technology, who purchased Photobit in 2001), Canon, Samsung, STMicroelectronics, Toshiba, OmniVision Technologies, Sony, and Foveon, among others. CMOS-type APS sensors are typically suited to applications in which packaging, power management, and on-chip processing are important. CMOS type sensors are widely used, from high-end digital photography down to mobile-phone cameras.

Advantages of CMOS compared with CCD

A big advantage of a CMOS sensor is that it is typically less expensive than a CCD sensor.

A CMOS sensor also typically has better control of blooming (that is, of bleeding of photo-charge from an over-exposed pixel into other nearby pixels).

In three-sensor camera systems that use separate sensors to resolve the red, green, and blue components of the image in conjunction with beam splitter prisms, the three CMOS sensors can be identical, whereas most splitter prisms require that one of the CCD sensors has to be a mirror image of the other two to read out the image in a compatible order. Unlike CCD sensors, CMOS sensors have the ability to reverse the addressing of the sensor elements. CMOS Sensors with a film speed of ISO 4 million exist.

Blooming in a CCD image.

Disadvantages of CMOS Compared with CCD

Since a CMOS sensor typically captures a row at a time within approximately 1/60th or 1/50th of a second (depending on refresh rate) it may result in a "rolling shutter" effect, where the image is skewed (tilted to the left or right, depending on the direction of camera or subject movement). For example, when tracking a car moving at high speed, the car will not be distorted but the background will appear to be tilted. A frame-transfer CCD sensor or "global shutter" CMOS sensor does not have this problem, instead captures the entire image at once into a frame store.

Distortion caused by a rolling shutter.

The active circuitry in CMOS pixels takes some area on the surface which is not light-sensitive, reducing the photon-detection efficiency of the device (back-illuminated sensors can mitigate this problem). But the frame-transfer CCD also has about half non-sensitive area for the frame store nodes, so the relative advantages depend on which types of sensors are being compared.

Architecture

Pixel

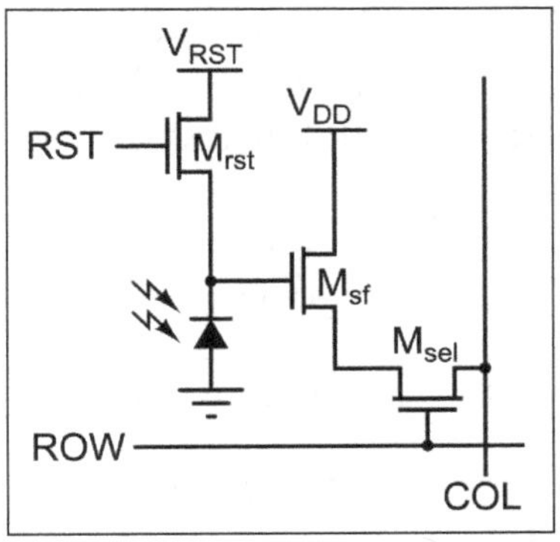

A three-transistor active pixel sensor.

The standard CMOS APS pixel today consists of a photodetector (a *pinned* photodiode), a floating diffusion, a transfer gate, reset gate, selection gate and source-follower readout transistor—the so-called 4T cell. The pinned photodiode was originally used in interline transfer CCDs due to its low dark current and good blue response, and when coupled with the transfer gate, allows complete charge transfer from the pinned photodiode to the floating diffusion (which is further connected to the gate of the read-out transistor) eliminating lag. The use of intrapixel charge transfer can offer lower noise by enabling the use of correlated double sampling (CDS). The Noble 3T pixel is still sometimes used since the fabrication requirements are less complex. The 3T pixel comprises the same elements as the 4T pixel except the transfer gate and the photodiode. The reset transistor, M_{rst}, acts as a switch to reset the floating diffusion to V_{RST}, which in this case is represented as the gate of the M_{sf} transistor. When the reset transistor is turned on, the photodiode is effectively connected to the power supply, V_{RST}, clearing all integrated charge. Since the reset transistor is n-type, the pixel operates in soft reset. The read-out transistor, M_{sf}, acts as a buffer (specifically, a source follower), an amplifier which allows the pixel voltage to be observed without removing the accumulated charge. Its power supply, V_{DD}, is typically tied to the power supply of the reset transistor V_{RST}. The select transistor, M_{sel}, allows a single row of the pixel array to be read by the read-out electronics. Other innovations of the pixels such as 5T and 6T pixels also exist. By adding extra transistors, functions such as global shutter, as opposed to the more common rolling shutter, are possible. In order to increase the pixel densities, shared-row, four-ways and eight-ways shared read out, and other architectures can be employed. A variant of the 3T active pixel is the Foveon X3 sensor invented by Dick Merrill. In this device, three photodiodes are stacked on top of each other using planar fabrication techniques, each photodiode having its own 3T circuit. Each successive layer acts as a filter for the layer below it shifting the spectrum of absorbed light in successive layers. By deconvolving the response of each layered detector, red, green, and blue signals can be reconstructed.

APS using Thin-film Transistors

For applications such as large-area digital X-ray imaging, thin-film transistors (TFTs) can also be used in APS architecture. However, because of the larger size and lower transconductance gain of TFTs compared with CMOS transistors, it is necessary to have fewer on-pixel TFTs to maintain image resolution and quality at an acceptable level. A two-transistor APS/PPS architecture has been shown to be promising for APS using amorphous silicon TFTs. In the two-transistor APS architecture on the right, T_{AMP} is used as a switched-amplifier integrating functions of both M_{sf} and M_{sel} in the three-transistor APS. This results in reduced transistor counts per pixel, as well as increased pixel transconductance gain. Here, C_{pix} is the pixel storage capacitance, and it is also used to capacitively couple the addressing pulse of the "Read" to the gate of T_{AMP} for ON-OFF switching. Such pixel readout circuits work best with low capacitance photoconductor detectors such as amorphous selenium.

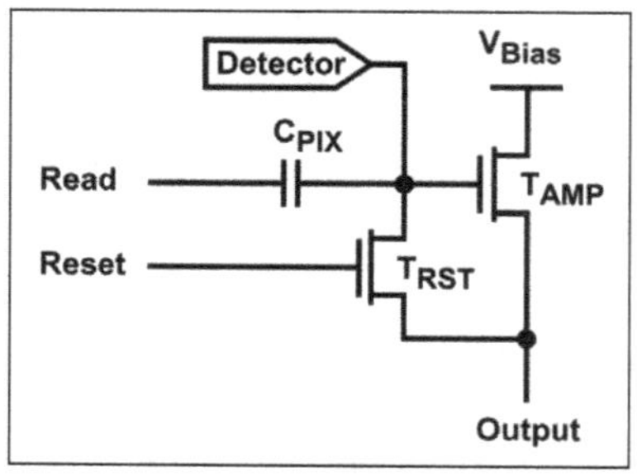

A two-transistor active/passive pixel sensor.

Array

A typical two-dimensional array of pixels is organized into rows and columns. Pixels in a given row share reset lines, so that a whole row is reset at a time. The row select lines of each pixel in a row are tied together as well. The outputs of each pixel in any given column are tied together. Since only one row is selected at a given time, no competition for the output line occurs. Further amplifier circuitry is typically on a column basis.

Size

The size of the pixel sensor is often given in height and width, but also in the optical format.

Design Variants

Many different pixel designs have been proposed and fabricated. The standard pixel is the most common because it uses the fewest wires and the fewest, most tightly packed transistors possible for an active pixel. It is important that the active circuitry in a pixel take up as little space as possible to allow more room for the photodetector. High transistor count hurts fill factor, that is, the percentage of the pixel area that is sensitive to light. Pixel size can be traded for desirable qualities such as noise reduction or reduced image lag. Noise is a measure of the accuracy with which the incident light can be measured. Lag occurs when traces of a previous frame remain in future frames, i.e. the pixel is not fully reset. The voltage noise variance in a soft-reset (gate-voltage regulated) pixel is $V_n^2 = kT/2C$, but image lag and fixed pattern noise may be problematic. In rms electrons, the noise is $N_e = \frac{\sqrt{kTC/2}}{q}$.

Hard Reset

Operating the pixel via hard reset results in a Johnson–Nyquist noise on the photodiode of or $V_n^2 = kT / C \ or \ N_e = \frac{\sqrt{kTC}}{q}$, but prevents image lag, sometimes a desirable tradeoff.

One way to use hard reset is replace M_{rst} with a p-type transistor and invert the polarity of the RST signal. The presence of the p-type device reduces fill factor, as extra space is required between p-and n-devices; it also removes the possibility of using the reset transistor as an overflow anti-blooming drain, which is a commonly exploited benefit of the n-type reset FET. Another way to achieve hard reset, with the n-type FET, is to lower the voltage of V_{RST} relative to the on-voltage of RST. This reduction may reduce headroom, or full-well charge capacity, but does not affect fill factor, unless V_{DD} is then routed on a separate wire with its original voltage.

Combinations of Hard and Soft Reset

Techniques such as flushed reset, pseudo-flash reset, and hard-to-soft reset combine soft and hard reset. The details of these methods differ, but the basic idea is the same. First, a hard reset is done, eliminating image lag. Next, a soft reset is done, causing a low noise reset without adding any lag. Pseudo-flash reset requires separating V_{RST} from V_{DD}, while the other two techniques add more complicated column circuitry. Specifically, pseudo-flash reset and hard-to-soft reset both add transistors between the pixel power supplies and the actual V_{DD}. The result is lower headroom, without affecting fill factor.

Active Reset

A more radical pixel design is the active-reset pixel. Active reset can result in much lower noise levels. The tradeoff is a complicated reset scheme, as well as either a much larger pixel or extra column-level circuitry.

Tachometer

Left figure shows a tachometer that can show up to 7000 RPM.

A tachometer (revolution-counter, tach, rev-counter, RPM gauge) is an instrument measuring the rotation speed of a shaft or disk, as in a motor or other machine. The device usually displays the revolutions per minute (RPM) on a calibrated analogue dial, but digital displays are increasingly common. Essentially the words tachometer and speedometer have identical meaning: a device that measures speed. It is by arbitrary convention that in the automotive world one is used for engine and the other for vehicle speed. In formal engineering nomenclature, more precise terms are used to distinguish the two.

In Automobiles, Trucks, Tractors and Aircraft

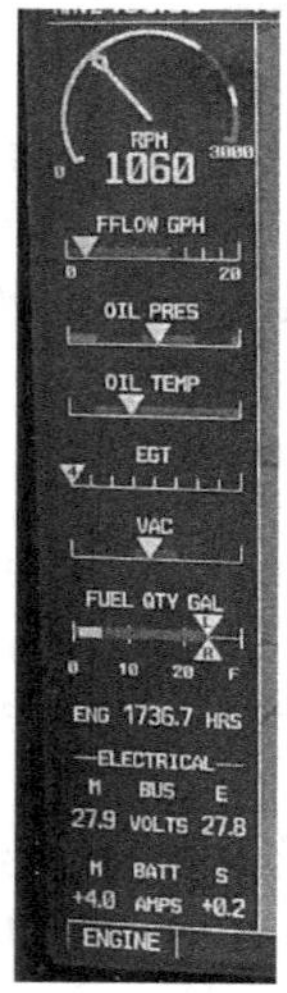

Cessna 172's G1000 tachometer (1,060 RPM) and engine hours (1736.7 hours).

Tachometers or revolution counters on cars, aircraft, and other vehicles show the rate of rotation of the engine's crankshaft, and typically have markings indicating a safe range of rotation speeds. This can assist the driver in selecting appropriate throttle and gear settings for the driving conditions. Prolonged use at high speeds may cause inadequate lubrication, overheating (exceeding capability of the cooling system), exceeding speed capability of sub-parts of the engine (for example spring retracted valves) thus causing excessive wear or permanent damage or failure of engines. This is more applicable to manual transmissions than to automatics. On analogue tachometers, speeds above maximum safe operating speed are typically indicated by an area of the gauge marked in red, giving rise to the expression of "redlining" an engine — revving the engine up to the maximum safe limit. The red zone is superfluous on most modern cars, since their engines typically have a revolution limiter which electronically limits engine speed to prevent damage. Diesel engines with traditional mechanical injector systems have an integral governor which prevents over-speeding the engine, so the tachometers in vehicles and machinery fitted with such engines sometimes lack a redline.

In vehicles such as tractors and trucks, the tachometer often has other markings, usually a green arc showing the speed range in which the engine produces maximum torque, which is of prime interest to operators of such vehicles. Tractors fitted with a power take-off (PTO) system have tachometers showing the engine speed needed to rotate the PTO at the standardized speed required by most PTO-driven implements. In many countries, tractors are required to have a speedometer for use on a road. To save fitting a second dial, the vehicle's tachometer is often marked with a second scale in units of speed. This scale is only accurate in a certain gear, but since many tractors only have one gear that is practical for use on-road, this is sufficient. Tractors with multiple 'road gears' often have tachometers with more than one speed scale. Aircraft tachometers have a green arc showing the engine's designed cruising speed range.

In older vehicles, the tachometer is driven by the RMS voltage waves from the low tension (LT) side of the ignition coil, while on others (and nearly all diesel engines, which have no ignition system) engine speed is determined by the frequency from the alternator tachometer output. This is from a special connection called an "AC tap" which is a connection to one of the stator's coil output, before the rectifier. Tachometers driven by a rotating cable from a drive unit fitted to the engine (usually on the camshaft) exist - usually on simple diesel-engined machinery with basic or no electrical systems. On recent EMS found on modern vehicles, the signal for the tachometer is usually generated from an ECU which derives the information from either the crankshaft or camshaft speed sensor.

Traffic Engineering

Tachometers are used to estimate traffic speed and volume (flow). A vehicle is equipped with the sensor and conducts "tach runs" which record the traffic data. These data are a substitute or complement to loop detector data. To get statistically significant results requires a high number of runs, and bias is introduced by the time of day, day of week, and the season. However, because of the expense, spacing (a lower density of loop detectors diminishes data accuracy), and relatively low reliability of loop detectors (often 30% or more are out of service at any given time), tach runs remain a common practice.

In Trains and Light Rail Vehicles

Speed sensing devices, termed variously "wheel impulse generators" (WIG), speed probes, or tachometers are used extensively in rail vehicles. Common types include opto-isolator slotted disk sensors and Hall effect sensors.

Hall effect sensors typically use a rotating target attached to a wheel, gearbox or motor. This target may contain magnets, or it may be a toothed wheel. The teeth on the wheel vary the flux density of a magnet inside the sensor head. The probe is mounted with its head a precise distance from the target wheel and detects the teeth or magnets passing its face. One problem with this system is that the necessary air gap between the target wheel and the sensor allows ferrous dust from the vehicle's underframe to build up on the probe or target, inhibiting its function.

Opto-isolator sensors are completely encased to prevent ingress from the outside environment. The only exposed parts are a sealed plug connector and a drive fork, which is attached to a slotted disk internally through a bearing and seal. The slotted disk is typically sandwiched between two circuit boards containing a photo-diode, photo-transistor, amplifier, and filtering circuits which produce a square wave pulse train output customized to the customers voltage and pulses per revolution requirements. These types of sensors typically provide 2 to 8 independent channels of output that can be sampled by other systems in the vehicle such as automatic train control systems and propulsion/braking controllers.

The sensors mounted around the circumference of the disk provide quadrature encoded outputs and thus allow the vehicle's computer to determine the direction of rotation of the wheel. This is a legal requirement in Switzerland to prevent *rollback* when starting from standstill. Strictly, such devices are not tachometers since they do not provide a direct reading of the rotational speed of the disk. The speed has to be derived externally by counting the number of pulses in a time period. It is difficult to prove conclusively that the vehicle is stationary, other than by waiting a certain time to ensure that no further pulses occur. This is one reason why there is often a time delay between the train stopping, as perceived by a passenger, and the doors being released. Slotted-disk devices are typical sensors used in odometer systems for rail vehicles, such as are required for train protection systems — notably the European Train Control System.

As well as speed sensing, these probes are often used to calculate distance travelled by multiplying wheel rotations by wheel circumference.

They can be used to automatically calibrate wheel diameter by comparing the number of rotations of each axle against a master wheel that has been measured manually. Since all wheels travel the same distance, the diameter of each wheel is proportional to its number of rotations compared to the master wheel. This calibration must be done while coasting at a fixed speed to eliminate the possibility of wheel slip/slide introducing errors into the calculation. Automatic calibration of this type is used to generate more accurate traction and braking signals, and to improve wheel slip detection.

A weakness of systems that rely on wheel rotation for tachometry and odometry is that the train wheels and the rails are very smooth and the friction between them is low, leading to high error rates if the wheels slip or slide. To compensate for this, secondary odometry inputs employ Doppler radar units beneath the train to measure speed independently.

In Analogue Audio Recording

In analogue audio recording, a tachometer is a device that measures the speed of audiotape as it passes across the head. On most audio tape recorders the tachometer (or simply "tach") is a relatively large spindle near the ERP head stack, isolated from the feed and take-up spindles by tension idlers.

On many recorders the tachometer spindle is connected by an axle to a rotating magnet that induces a changing magnetic field upon a Hall effect transistor. Other systems connect the spindle to a stroboscope, which alternates light and dark upon a photodiode.

The tape recorder's drive electronics use signals from the tachometer to ensure that the tape is played at the proper speed. The signal is compared to a reference signal (either a quartz crystal or alternating current from the mains). The comparison of the two frequencies drives the speed of the tape transport. When the tach signal and the reference signal match, the tape transport is said to be "at speed." (To this day on film sets, the director calls "Roll sound!" and the sound man replies "Sound speed!" This is a vestige of the days when recording devices required several seconds to reach a regulated speed).

Having perfectly regulated tape speed is important because the human ear is very sensitive to changes in pitch, particularly sudden ones, and without a self-regulating system to control the speed of tape across the head, the pitch could drift several percent. This effect is called a wow-and-flutter, and a modern, tachometer-regulated cassette deck has a wow-and-flutter of 0.07%.

Tachometers are acceptable for high-fidelity sound playback, but not for recording in synchronization with a movie camera. For such purposes, special recorders that record pilottone must be used.

Tachometer signals can be used to synchronize several tape machines together, but only if in addition to the tach signal, a directional signal is transmitted, to tell slave machines in which direction the master is moving.

Air Pollution Sensor

Air pollution sensors are devices that detect and monitor the presence of air pollution in the surrounding area. They can be used for both indoor and outdoor environments. These sensors can be built at home, or bought from certain manufactures. Although there are various types of air pollution sensors, and some are specialized in certain aspects, the majority focuses on five components: ozone, particulate matter, carbon monoxide, sulfur dioxide, and nitrous oxide. The sensors were very expensive in the past, but with technological advancements these sensors are becoming more affordable and more widespread throughout the population. These sensors can help serve many purposes and help bring attention to environmental issues beyond the scope of the human eye.

The EPA maintains a repository of air quality data through the Air Quality System (AQS), where it stores data from over 10,000 monitors in the United States.

While use of these sensors was expensive in the past, the 2010s saw a recent trend towards the development of cheaper portable air-quality sensors that can be worn by individuals to monitor local air quality levels. These sensors, can then, in turn, help measure the spatiotemporal coverage and variety of chemical species, and empower individuals and communities to better understand their exposure environments and risks from air pollution.

A research group led by William Griswold at UCSD handed out portable air pollution sensors to 16 commuters, and found "urban valleys" where buildings trapped pollution. The group also found that passengers in buses have higher exposures compared to those in automobiles.

Health

Scientific evidence has indicated that indoor air pollution can be worse than outdoor pollutants in large and industrialized cities. Many products and chemicals used inside the home, for cooking and heating, and for appliances and home décor are primary sources of indoor air pollutants. Everything we use in the home contributes to the pollution, and can possibly degrade the environment. Air pollution is responsible for 7 million premature deaths around the world each year. When pollutants enter the body through our respiratory system, they can be absorbed in the blood and travel throughout the body, and can directly damage the heart and other vital organs.

Health Canada's research confirms that poor air quality is hazardous to human health and as the population ages more Canadians are at risk. Young children are at risk because on a per-body weight basis, they tend to inhale relatively more air than adults, and their immature defence systems make them more susceptible to air pollution. The elderly are also at risk due to their weaker lungs, heart, and defence systems. The elderly also tend to have undiagnosed respiratory or cardiovascular health conditions. People participating in strenuous work or sporting activities outdoors breathe more deeply and rapidly than their sedentary counterparts. This increases the risk of these individuals developing air pollution related diseases. Due to its aging and expanding population, the number of Canadians affected grows each year. It is estimated that tens of thousands of Canadians die prematurely each year due to acute air pollution and that high summer temperatures lead to increased illnesses, hospitalization and deaths among older adults.

References

- Humidity-Sensors-A-Review, publication: researchgate.net, Retrieved 13 July, 2019
- Goble, Steve (2003). "Rebreathers". South Pacific Underwater Medicine Society Journal. 33 (2): 98–102. Retrieved 2009-03-20
- What-is-a-biosensor-types-of-biosensors-and-applications: elprocus.com, Retrieved 14 August, 2019
- Gaikwad, Pramod. "Galvanic Cell Type Sensor for Soil Moisture Analysis". Analytical Chemistry. 87: 7439–7445. Doi:10.1021/acs.analchem.5b01653
- Hall-effect, electromagnetism: electronics-tutorials.ws, Retrieved 15 January, 2019
- A review of the pinned photodiode for CCD and CMOS image sensors, IEEE J. Electron Devices Society, vol 2(3) pp. 33-43 May 2014 open access "Archived copy". Archived from the original on 2015-10-27. Retrieved 2014-08-17

Applications of Sensors

4

- **Sensors in Medical Science**
- **Industrial Applications of Sensors**
- **Sensors Applications in Transport and Logistics**
- **Remote Sensing**
- **Collocation**

Sensors are applied in a number of fields including medical science, transport and logistics, remote sensing, collocation, etc. This chapter has been carefully written to provide an easy understanding of these applications of sensors as well as the use of optical chemical sensors in industries.

Sensors in Medical Science

Monitoring the use of Wearables

Parameters

Advancements in sensor technologies have resulted in a variety of wearable computers and sensors that can be used in either contact or noncontact mode for biomedical applications. Several wearable devices are commercially available for monitoring patients' vital signs on a continual basis, which is crucial for modern digital health, and they allow for data transfer between the device and a smartphone or other devices (e.g., those at healthcare units) through Bluetooth, Wi-Fi, or other wireless technologies. These devices not only monitor vital signs, but they also measure other parameters from the daily activities of patients to aid in their health status, such as calories burned, step monitoring, sleep patterns, activity levels, and eating habits. In particular, heart rate, blood oxygen saturation level, body and/or skin temperature, respiration rate, and blood pressure are the major physiological parameters that can be monitored using the majority of wearables.

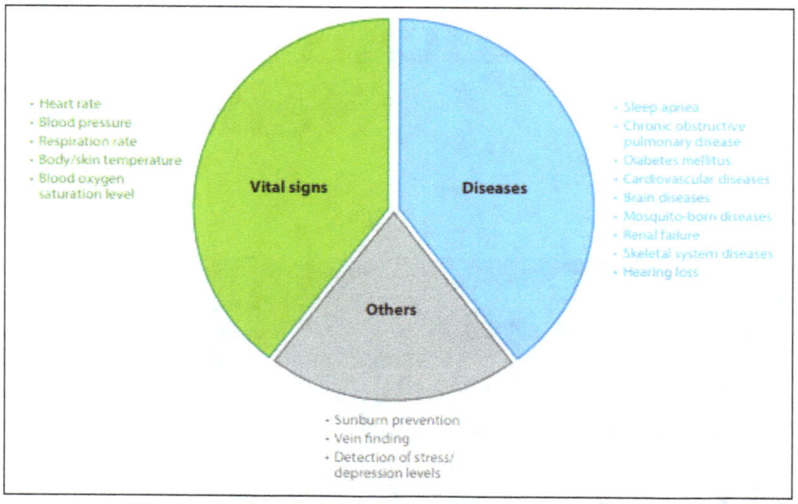

Some biomedical applications of wearables, including those to monitor vital signs, a variety of diseases, and other factors.

Heart rate is a commonly measured parameter using wearables, and this vital sign can easily be detected from various body parts. The first wireless heart rate wearable, a wristband, was introduced in the 1980s to monitor the real-time performance of athletes during exercise. Given the popularity of continuous heart rate monitoring, which is particularly important for fitness-oriented individuals, dozens of wearable models are currently available in the market that can measure the heartbeat rate at rest, as well as during walking and running, from various places on the body, including the ear, wrist, finger, and chest, by means of optical sensors, gyroscopes, accelerometers, and pressure sensors embedded in the wearables. These devices not only provide the heartbeat rate of users but also time-based plots and recordings of the history together with global positioning systems (GPS) for improved performance analysis.

Another vital sign measured by wearables is blood oxygen saturation level, which is a measure of breathing and circulation of blood in the vessels. The normal range for the arterial oxygen level is 75–100 mm Hg, while the range lower than 60 mm Hg indicates the need for supplemental oxygen. It can be measured noninvasively from the ear and fingertip using pulse oximeters integrated into wearables operating with light-emitting diodes (LEDs) at two wavelengths at least, where one is less than 810 nm (e.g., 660 nm) and the other is higher than 810 nm (e.g., 1,000 nm). The sequential sampling of light transmitted through the finger or ear is carried out using a photodiode lying on the opposite side of the tissue, which enables determination of the blood oxygen saturation level through a ratiometric analysis of the oxygenated and deoxygenated blood hemoglobin in the presence of a pre-established calibration curve.

Table: Physiological Parameters and Examples of Commercial Wearable Devices and their Sensors.

Vital sign	Wearable device	Sensor
Heart rate	Earring (e.g., Ear-O-Smart)	Pulse oximeter
Heart rate	Earpiece (e.g., Cosinuss, one earpiece)	Pulse oximeter
Heart rate	Earphone (Bragi)	Pulse oximeter
Heart rate	Watch (e.g., Samsung Gear Fit, Basis)	Pulse oximeter

Heart rate	Fingertip (e.g., Go2 Fingertip Pulse Oximeter)	Pulse oximeter
Heart rate	Sensor patch (Avery Dennison Metria)	Electrocardiogram
Heart rate	Wristband (e.g., Phyode W/Me wristband) Wristband (e.g., iHealth, Omron)	Touch-activated electrocardiogram sensor Pressure sensor
Heart rate	Glass (e.g., Google Glass)	Gyroscope, accelerometer, image sensor
Heart rate	Shirt (e.g., PoloTech)	3D accelerometer
Blood oxygen saturation level	Earphone (Bragi)	Pulse oximeter
Blood oxygen saturation level	Fingertip (e.g., Go2 Fingertip Pulse Oximeter)	Pulse oximeter
Body and/or skin temperature	Earpiece (e.g., Cosinuss one earpiece)	Temperature sensor
Body and/or skin temperature	Tattoo (e.g., VivaLnk Fever Scout)	Temperature sensor
Body and/or skin temperature	Earphone (Bragi)	Temperature sensor
Body and/or skin temperature	Watch (e.g., Basis)	Temperature sensor
Respiration rate	Belt accessory (e.g., Spire)	Accelerometer
Respiration rate	Smart textile electrocardiogram sensor	Piezo-resistive sensor
Respiration rate	Glass	Gyroscope, accelerometer, image sensor
Blood pressure	Sensor	Two photoplethysmography sensors
Blood pressure	Patch	Potential difference
Blood pressure	Wristband (e.g., iHealth, Omron)	Pressure sensor

Another vital sign of interest for wearables is the body temperature, which is within the range of 36.5 °C (97.8 °F) to 37.2 °C (99 °F) for a healthy adult. An increase in body temperature is an important sign that is mostly related to disturbance in the immunological system of patients. It can be measured from the mouth, rectum, eardrum, under the arm, and through the skin. Because the skin temperature is easily affected by external factors, and the temperature control unit of the body is located near the hypothalamus, it is preferable to measure body (core) temperature rather than the skin temperature, which is done from places close to the innermost organs. Unlike heart rate, body temperature is not measured continuously, but at certain time intervals, because the change in body temperature is a slow process. Patches, earpieces, and watches are typical wearables for measuring body temperature.

Breathing rhythm or respiration rate monitoring is one of the basic methods for determining the health status of a patient when performing clinical evaluations. Continuous monitoring of respiration rate can provide information on the progression of an illness such as cardiac arrest. There are two modes for measuring breathing rhythm: (a) contact mode, by means of attaching sensing devices to the bodies of patients, and (b) noncontact mode, by capturing images of the patients using infrared-based sensors. Patches, bras, shirts, and bracelets are some examples of wearable devices for monitoring respiration rate.

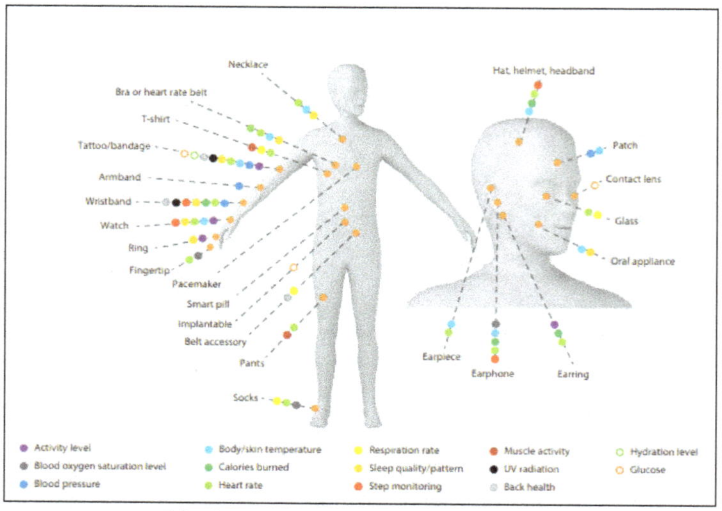

Wearable devices worn on various body parts and the parameters that they can monitor.

Another vital sign is blood pressure, which refers to the pressure of blood in the circulatory system; a pressure less than 120/80 mm Hg is considered normal by clinicians. If the values are +10 higher than normal, this means the patient has prehypertension and needs to begin a treatment program to regain healthy blood pressure levels. High blood pressure levels can be an indicator of damage to artery walls. Therefore, monitoring blood pressure levels can be useful in the prevention and management of high blood pressure. Wearable monitoring devices such as wrist-and arm-based blood pressure monitors are based on noninvasive measurements with skin contact, while avoiding errors due to the movement of patients during measurement.

Diseases and Disabilities

Mobile technologies are paving the way for revolutionizing the diagnosis of diseases and monitoring patient treatment. Because wearables are personal devices that provide objective and continuous monitoring of vital parameters, they can be used as a clinical tool to determine the patterns of a particular disease, provide an improved understanding of the disease, and monitor the health information of users. Although wearables currently on the market are mainly used for general fitness and wellness purposes, various devices are being designed for making a difference in the lives of people by gathering information on a particular disease. In this topic, we discuss some of these examples in which wearable technologies are used.

Sleep apnea: This extremely common condition results in interruptions or a decrease in breathing for a few seconds up to a minute. It affects approximately 18 million people in the United States alone. The treatment type depends on the severity of the case and ranges from weight loss to surgical operations. DentiTrac is an FDA-cleared wearable oral device for following the prescribed therapy for sleep apnea. It measures the temperature, movement, and head position of patients by determining the spatial orientation of the device in the mouth.

Chronic obstructive pulmonary disease: Chronic obstructive pulmonary disease (COPD) is another common lung disease that leads to shortness of breath. It affects 210 million people worldwide and is a major health problem for people 40 years and older. Various sensors, including an ear wearable, have been developed to monitor the physical activities of patients with COPD. These wearables allow patients to continuously evaluate their condition at home, where they are away from most of the air pollutants and bioaerosols associated with healthcare units. Therefore, using wearables for COPD may reduce healthcare costs for patients that can be treated at home.

Diabetes mellitus: This is a chronic disease whereby the body cannot produce sufficient insulin, and the control of blood glucose levels is essential for diabetic patients. A wearable artificial pancreas is commercially available for glucose level monitoring and regulation. It is composed of a flexible core system as a brain and three wedges for insulin delivery, glucose sensing, and glucagon delivery. Another wearable that is being designed to measure blood sugar levels in diabetics is the smart contact lens that Google/Verily Life Sciences has been developing in collaboration with Novartis (Basel, Switzerland). However, there is very little public domain information about the technical details and feasibility of this approach.

Cardiovascular diseases: This group of diseases is related to the heart and veins, including venous thrombosis, heart failure, and cardiac dysrhythmia. According to a World Health Organization report, more than 17 million patients died from cardiovascular diseases in 2008, and 23 million people are threatened by these diseases each year. Various wearable sensors exist for providing real-time heart rate measurements, such as the wireless blood pressure wrist monitor by iHealth, which monitors blood pressure in connection with a smartphone. Validation of the performance of this wrist monitor was investigated by Wang et al., and it was shown that the accuracy of the measurements was in good agreement with the reference clinical measurements. The standard deviation between the two methods was found to be within ± 7 mm Hg for both systolic blood pressure and diastolic blood pressure.

Brain diseases: Dementia refers to a group of brain diseases leading to a decrease in cognitive abilities. This group includes Alzheimer's, Huntington's, and Parkinson's diseases as well as vascular dementia. The Vega GPS bracelet is a wearable sensor for ensuring the safety of people with Alzheimer's disease by monitoring their location with the use of GPS and global system for mobile communications positioning. Epilepsy is another group of brain disorders that causes seizures. Embrace by Empatica (Milan, Italy) is a wristband for monitoring physiological signals in epileptic people in real time to alert others, for example, family members.

Mosquito-borne diseases: Mosquitoes cause a wide range of deadly diseases, such as malaria, chikungunya, yellow fever, the Zika virus, and the Ebola virus, which affect millions of people worldwide. The Kite Patch is a patch-type wearable that disperses volatile compounds and is worn on a shirt to repel mosquitoes. TermoTell is a bracelet for monitoring the temperature and sweat patterns of a child and sends a notification to the smartphones of its parents if there is a match on the data gathered with malaria symptoms.

Renal failure: This condition is also known as kidney failure and can be categorized as acute kidney injury and chronic kidney disease. In the treatment of renal failure, dialysis is commonly used, in which kidney function is replaced by a machine. To replace dialysis, a wearable artificial kidney has been developed, and its current version weighs only ~4.5 kg.

Skeletal system diseases: Joint disorders, osteoporosis, and poor posture are among the most common skeletal conditions. Using three-dimensional (3D) gyroscopes, accelerometers, and magnetometers embedded into wearable sensors (e.g., CUR Smart Pain Relief and Valedo), the chronic pain resulting from most skeletal diseases can be treated with transcutaneous electrical nerve stimulation and by performing therapeutic exercises. Another wearable, Lumo Lift, monitors postural variation and warns users through vibrations when they deviate from normal posture, reminding them to return to a normal posture.

Sunburn prevention: The ultraviolet (UV) radiation of sunlight has harmful effects on the skin, causing wrinkles, burns, aging, and even skin cancer. Approximately three million people per year are diagnosed with skin cancer globally. Wearable UV sensors, which can be worn on the arm in the form of a bracelet, armband, or wristband, can be used to monitor UV exposure levels with alerts for potential skin damage and safety precautions, as well as estimating vitamin D production levels.

Vein ftnding: A wearable smart glass termed Eyes-On technology enables nurses to rapidly see the veins of patients through the skin by incorporating multispectral 3D imaging and wireless connectivity.

Detection of stress/depression levels: Wearables can also be used to determine the state of mind of their users. For example, Airo Health's (Canada) recent product is a wristband that monitors heart rate variability and aims to warn the user about a rise in personal stress levels.

Wearable Technologies for different Body Parts

Wearable sensors can be worn on the head, body, arm, and leg/foot, or they can be implantable, such as smart pills or artificial organs. The common purpose of all wearable sensors is to provide useful data while the user is wearing the sensors and performing activities such as walking, sleeping, and eating. Most of these sensors record or transfer the user data to mobile devices or smartphones through dedicated applications. Starting with the next subsection, we summarize some examples of different body parts for which wearables have been designed.

Head

The head is the uppermost part of the body and includes the forehead, ears, eyes, nose, and mouth. The wearable health-tracking devices for the head comprise mainly glasses, goggles, contact lenses, hats, headbands, hearing aids, earrings, earphones, and patches.

Modern smart glasses can be considered as wearable computers, and they are embedded with several sensors, such as gyroscopes, accelerometers, pressure sensors, image sensors, and microphones. They can be operated using voice commands through a user interface. Google Glass, Recon Jet (Recon Instruments; Vancouver, BC), and JINS MEME (Jin-Co Ltd., Japan) are some examples of wearable glasses. The video and audio capturing abilities of smart glasses can be used to record and archive educational videos for biomedical experts: for example, how to perform surgical operations or in general for training and educating new professionals. Moreover, smart glasses can be used to monitor physiological parameters such as heart rate and respiratory rate and as a rapid diagnostic test reader for quantifying the intensity levels of a test line for different target analyte concentrations. As another example, a wearable glass platform for healthcare providers, introduced by Evena Medical (Roseville, CA), provides clinicians with an easy means of finding patients' veins using infrared sensors.

Smart contact lenses are contact lenses embedded with multisensors (e.g., pressure and temperature sensors). They are similar to implants but do not require any surgical operation and can be easily worn or removed by the user. Possible biomedical uses of smart lenses include detecting the glucose levels in tears, improving vision without glasses by focusing the eye instantly, and monitoring the progression of glaucoma in patients by constantly measuring the curvature of the eye lens.

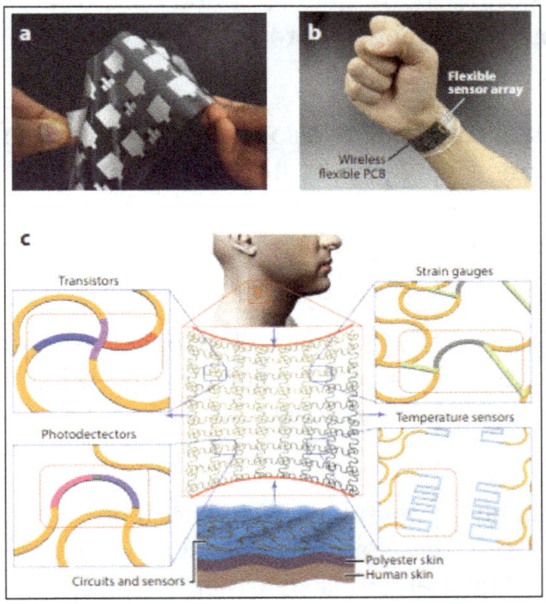

Examples of wearable sensors: (a) Skin-worn wearable for real-time monitoring of lactate and electrocardiogram using electrochemical sensor and bipolar electrodes. (b) Smart wristband for multiplexed perspiration analysis. (c) Flexible skin.

Hats, helmets, and headbands can be also embedded with wearable sensors. The smart hat and smart headband introduced by Spree Wearables, Inc. (Dallas, TX) and the LifeBEAM Helmet monitor the parameters commonly tracked by other wearables, such as heart rate and body temperature, using a built-in advanced plethysmograph, as well as optical and other sensors. Other than activity monitoring, wearables worn on the head can be used to measure brain activity. For example, SmartCap, introduced by SmartCap Technologies (Australia) is a fatigue monitoring system that measures brainwave signals for the risk of microsleep, which is an unintended/uncontrolled sleep for a duration of 5–10 s. The Melon headband and IMEC electroencephalographic (EEG) headset monitor the brain's EEG signals to measure mental activity for clinical and research applications, such as for the estimation or quantification of a user's focus levels.

Wearables worn on the ear include hearing aids, earbuds, earrings, and earphones that monitor physiological parameters using pulse oximetry. The X-Patch (X2 Biosystems Inc.; Redwood City, CA) is another wearable device worn on the back side of the ear that monitors potential head injuries, which are important for people taking part in sporting activities, and for military personnel and industrial workers.

Certain wearables are worn inside the mouth. Examples include a mouthguard for detecting different levels of uric acid or lactate in the saliva and DentiTrac, which is an oral device worn on any oral appliance and used to monitor sleep apnea and adherence of patients to sleep therapy.

Torso

The torso is the central part of the body in which most of the organs are housed. Suits, belts, and underwear are some examples of wearable technologies worn on the torso.

Some examples of wearable technologies worn as a suit include pajamas for electrocardiogram measurement in babies, baby glove swaddles for temperature sensing, smart jackets for monitoring the physiological parameters of newborn infants, military uniforms with a wearable computer to continuously monitor military personnel and protect them, e.g., from environmental threats, and swimsuits with a UV sensor.

Patches or tattoos with wireless connectivity also hold great promise for the continuous monitoring of vital parameters and improving patient health. The compatibility of patches with several key applications have been demonstrated, such as a tattoo for measuring the skin temperature of a baby, a plaster for sweat analysis , and patches for drug delivery, pain relief, and repelling mosquitoes.

Smart fabrics, which are textiles embedded with electronics, are also types of wearable sensors worn on the torso. A few examples include the smart bra and shirts for monitoring physiological parameters such as heart rate and respiration rate, as well as the chest belt as a sensor for monitoring the respiration rate.

Artificial organs are also available as wearable systems. A good example is the wearable pancreas, Genesis, introduced by Pancreum, Inc. (San Francisco, CA). It is used for measuring and regulating glucose levels in the blood together with a smartphone application.

Arm, Leg and Foot

Wearable devices worn on the arm, leg, and foot are mostly accessories such as smart watches, bracelets, rings, armbands, and wristbands that can monitor physiological parameters such as body temperature and heart rate, UV exposure levels, and daily activities. Smart socks and sleeves for injury prevention are a few of the wearable accessories worn on the foot or leg. Smart socks, introduced by Alpha-Fit GmbH (Wertheim, Germany), provide 3D dynamic measurement of pressure distribution over the entire foot while a patient is walking. This foot pressure distribution serves as valuable data for orthopedic shoemakers in fabricating custom shoes for diabetic foot syndrome.

Ingestible and Implantable Wearables

Ingestible pills, introduced by Proteus Digital Health (Redwood City, CA), are smart pills for monitoring the precise time at which medicine is taken. When the smart pill reaches the stomach, it is powered up with the chemical reaction with the stomach fluid and sends an ingestion time signal to the patch worn on the body. This patch not only communicates with the smart pill but also monitors heart rate, blood pressure, pH, and temperature.

Another example of implantable wearables is a pacemaker that is used to treat irregular heartbeats, known as arrhythmias. Once it detects an irregular heartbeat, it provides low-energy electrical pulses to restore the normal rhythm. The ICD is a newer version of a pacemaker, which operates in the same manner. However, if the heartbeat cannot be restored to a normal rhythm, the ICD provides high-energy electrical pulses.

Another recent example of implantable wearables is a pulse generator, which is used for deep brain stimulation. It is a battery-operated device with a similar size to a stopwatch. It is implanted in the brain by means of surgery and provides electrical signals to control movement, particularly for patients with Parkinson's disease.

Materials used in Fabrication of Wearables

Wearables are worn mostly on the soft and curved parts of the human body and require flexible electrodes or sensors to continuously measure physiological parameters without losing the precision and accuracy of the data obtained during their use.

Smart textiles and stretchable electronics are two emerging material technology fields that offer considerable promise for fabricating thin, flexible, washable, soft, and durable electronics for the next generation of wearables.

Smart Textiles

Rapid advancements in material technology have enabled the fabrication of electronics that are embedded into fabrics or cloth, resulting in various biomedical applications. One example is biometric shirts, which can be used by civilians or military personnel for a variety of purposes such as environmental sensing and monitoring of physiological parameters. The integration of electronics with textiles can be carried out in two ways: (a) embedding them into cloth and (b) integrating them into the textile fibers, which are flexible and thin materials with a large height-to-length ratio. The former method requires textile fibers made of electrically conductive materials, obtained by embedding or twisting conductive powders, particles, fillers, or yarns (e.g., silver, copper or carbon) into nonconductive fabric fibers (e.g., Kevlar), or using inherently conductive polymeric fibers (e.g., polyaniline and polpyrrole).

The substrate of the conductive fiber, which is the nonconductive part of the material, provides the desired strength or flexibility, while the embedded conducting elements behave as electrodes on the fabric or yarn. These two parts of the fiber exhibit different physical properties. For example, the resistance of a natural fiber such as wool is affected by air humidity, whereas the electrical resistance of a metal fiber changes with temperature. Although the substrate provides strength and flexibility, any damage to the conductive fibers can lead to the loss of electronic functionality, which is an important property for sensing applications. However, these types of failures can be avoided using a large number of fibers on the surface area. Another method of embedding electronics into textiles, which is compatible with batch processing, is the encapsulation of the electronics into a bundle of fibers. Using this method, the chip is protected from chemical, mechanical, and thermal stresses and deformations (e.g., day-to-day wearing of the textile-based material) by impregnating a polymer resin between the fiber bundle and electronics. The polymer resin acts like a shield to protect the chip from external effects (such as liquid). Radio frequency identification chips, LEDs, and thermistors are examples of electronic products that can be embedded into yarns. Smart textiles are used for monitoring the physiological signals of the human body or to control devices wirelessly using interfaces.

The Adidas miCoach seamless sports bra and racer tank are made of conductive fibers that are knitted into the textile to sense heart pulses. The electronic functionality of the wearable sensor is achieved with a module attached to the bra, by means of which the gathered data are transmitted to the receiver, such as a smartphone or smart watch. Another commercially available wearable that uses smart textiles is the crewneck smart shirt known as PoloTech, by Ralph Lauren. This machine-washable shirt is made of a fabric composed of 70% polyester, 21% nylon, and 9% spandex, embedded with silver fibers and integrated with a 3D accelerometer. The data collected using the silver fibers are transmitted to an iPhone or iPod via Bluetooth using a USB-charged black box attached to the shirt with five pins; it also monitors heart rate and other daily activities. Furthermore, smart textiles have also been used for pressure-sensing applications. For example, the smart sock, briefly introduced earlier, is a wearable sensor that provides data regarding the foot structure of patients for the production of custom-fit shoes. These orthopedic shoes help patients with diabetic foot syndrome by reducing foot pain and the risk of developing foot ulcers.

Smart textiles, in the field of wearable computers or sensors, offer various advantages, including flexibility and multifunctionality. However, several design difficulties still need to be addressed to ensure continuity of their functionalities for selected applications in terms of reliability and stability. Comfort, cost-effectiveness, selection of the sensor positions for different body sizes, the batch production complexity of the integration of electronics into textiles, reusability of smart textile products, and energy consumption are some of the most important issues in these materials that need more research and development.

Stretchable Electronics

The materials used in the design of wearable devices should be flexible, lightweight, thin, and conformable to effectively facilitate personal health monitoring. To this end, stretchable electronics present opportunities for bending and stretching of electronic circuits and sensors together with their surroundings. In this field of research, various materials (e.g., paper, leather, and plastic films, particularly polymers) have been used as a substrate to keep the circuit together while achieving the required flexibility. Polyethylene terephthalate, polyimide (Kapton), polycarbonate, polyethylene naphthalate, CYTOP, poly(methyl methacrylate), polystyrene, and polydimethylsiloxane (PDMS) are polymers commonly used as a protective dielectric layer for wearable patches. Among these, PDMS is mostly used as a dielectric support material owing to its ease of preparation, transparency, stiffness, and biocompatibility. Although PDMS is an effective material for rapid prototyping, it results in adhesion problems on nonpolymeric substrates, such as metal surfaces, and requires surface treatments (e.g., oxide coating or ozone treatment) to facilitate the adhesion of PDMS to other surfaces. There are certain other drawbacks to using polymers in wearable technology. For example, they are not suitable for processing steps requiring high temperatures, such as above 300 °C, and their mechanical properties change when they are exposed to temperature and humidity. Moreover, plastic films may allow gas (such as oxygen or water vapor) diffusion through the thickness of the sheets, resulting in changes in the chemical structure of the electrodes, such as oxidation of the metal layers. If there is no need for transparent materials, these drawbacks can be eliminated by using other materials such as thin metal layers as a substrate, which can be processed at high temperatures and do not allow the permeation of oxygen and water vapor.

In addition to the substrate, electrodes should also be flexible because the compatibility of electronic circuits with these flexible substrates is important for fully functional devices following long-term use. Various materials can be used to fabricate stretchable electrodes, such as conductive inks, graphene sheets, nanowires, nanoribbons, and nanotube films with different design structures, including lines (e.g., Peano), loops (e.g., Moore), and branch-like meshes (e.g., filamentary serpentine and Greek cross). Each of these has a different stress-strain response that can be readily applicable to wearable sensors. The sensor-containing stretchable electronics should be in close contact with the body because low impedance between the body and the electrode surface is important for improving the performance of the medical device and providing a high signal-to-noise ratio.

Stretchable electronics have been used in wearable patches, particularly for skin-mounted sensors or electrodes that need to follow the human body shape to monitor physiological parameters in the heart, brain, and muscles. For example, Blue Spark Technologies introduced a patch thermometer for measuring the skin temperature of a child. This wearable sensor contains a printed flexible battery, continuously senses the temperature, and sends the data to a mobile phone by means of a smart application.

Standards in Wearable Devices

Wireless Medical Devices

Wearables are generally used to monitor at least one physiological sign and perform at least one function based on wireless radio frequency communication, such as transferring collected patient data to a source by means of a mobile application. There are several advantages to using wireless technology in wearable computers or sensors, including compactness, mobility, allowing doctors remote access to patient data, and providing real-time monitoring of patient data. Although wearable sensor developers in the industry refer to particular standards and guidelines that are specific to their devices, there are several other FDA-recognized standards that exist for wireless medical devices that should be taken into consideration during the design, testing, and use of these wearable devices. The International Organization of Standardization (ISO), International Electrotechnical Commission (IEC), Institute of Electrical and Electronics Engineers (IEEE), American National Standards Institute (ANSI), and Association for the Advancement of Medical Instrumentation (AAMI) are other major organizations that provide such standards.

During the design of medical devices, consideration should be given to the selection of wireless technology, e.g., wireless medical telemetry service, IEEE 802.11 or Bluetooth, wireless connection quality, wireless security, and operation frequency with minimum interference in the operation band of other nearby radio frequency devices, not only in the country in which the product is designed, but also in other countries where it will be used. Failure causes and the associated risks and performance of the technology should be evaluated under low wireless connectivity using certain parameters, such as signal-to-noise ratio, to prevent any loss during data transfer and storage. Because wireless medical devices are associated with radio frequency technology, the properties of the intended use of the medical devices should match the expected performance of the wireless technology. Therefore, it is crucial to consider the functions that will operate wire- lessly. For this purpose, ISO 14971, a standard for the estimation, identification, control, and monitoring of risks associated with medical devices, has been used to assess the performance of the devices during the product life cycle.

Various devices operate at low-frequency bands (<450 MHz), such as television and metal detectors, and high-frequency bands (450–3,000 MHz), such as cellular phones, whereas most medical devices operate at frequencies ~2,400 MHz. In recent years, there has been a tremendous increase in the number of wireless device users, and with developing technology, the size of these devices has been reduced considerably, so that patients may carry these wireless consumer devices near an implanted medical device while being close to its operating frequency. As a result, the electromagnetic fields that wireless medical devices are exposed to are not the same as prior to 20 years ago. Therefore, it is essential to create an electromagnetically compatible medical device to minimize interference with other devices in complex electric and magnetic fields. ISO 14117 (2012) is a standard that specifies procedures (such as test setup details, test procedure, and performance criteria) for assessing the electromagnetic compatibility of active implantable medical devices used for heart rate irregularity therapies, such as pacemakers and ICDs. This standard also provides measures for protecting medical devices from external static or AC magnetic fields faced during usage. The devices that comply with the rules of this standard are designed to have certain specifications, such as device size, sensing limits, and battery life. Within the scope of the standard, the devices are tested for frequency limits of up to 3,000 MHz.

IEC 60601-1-2 is another FDA-recognized technical standard that specifies tests and requirements for the electromagnetic compatibility of wireless medical devices. Two national standards on the electromagnetic compatibility of wireless medical devices are available in the United States: AAMI TIR18 and ANSI C63.18. AAMI TIR18, provided by the US Association for the Advancement of Medical Instrumentation, is a report that specifies recommendations regarding the use of wireless medical devices in healthcare associations, including identification of all radio frequency sources and managing electromagnetic fields to ensure the safety of patients. This report also provides very specific guidance on the management of electromagnetic fields and specifies recommendations for the use of wireless medical devices, not only at existing healthcare units, but also for new facilities. These guidelines are for design considerations as well as the placement and shielding of electromagnetic devices to minimize patients' field exposure. ANSI C63.18, on the other hand, provides a test methodology for clinical and biomedical engineers to estimate the radiation immunity of medical devices to portable radio frequency transmitters.

ISO/IEEE 11073 is a family of standards for medical devices and has a subgroup for personal health devices (PHDs). These PHD standards are based on device specialization and are available for such devices as blood pressure monitors, pulse oximeters, glucose meters, thermometers, weighing scales, and cardiovascular fitness and activity monitors, among others. These standards are based on the operation of two autonomously communicating systems: agents as small inexpensive devices such as a wearable necklace and managers in the form of computational devices such as smartphones. The manager can control the agent and send certain tasks to it, while the agent transfers the collected data to the manager for data processing. The analysis is then shown to the user and/or remotely sent to the healthcare provider.

Mobile Applications

The FDA defines a mobile application as "a software application that can be executed (run) on a mobile platform (i.e., a handheld commercial off-the-shelf computing platform, with or without wireless connectivity), m or a web-based software application that is tailored to a mobile platform but is executed on a server". If a mobile application is used as an attachment to a regulated medical device or transforms a mobile device such as a smartphone or tablet into a regulated medical device, it is referred to as a mobile medical application. The FDA regulates mobile applications using a risk-based classification approach, where each classification has different requirements, such as premarket approval. A mobile medical application falls under the regulatory oversight of the FDA because it may present risks to the health of users. On the other hand, a regular mobile application is not a medical device and therefore does not fall under the regulatory oversight of the FDA, as it does not constitute risks to the health of users, for example, if it is developed and intended to be used by clinicians as an educational tool. Another classification also exists for mobile applications, for which the FDA "intends to exercise enforcement discretion." This is because such applications pose a low risk to the health of users, for example, if the application is for carrying out simple calculations commonly used by healthcare providers, such as body mass index calculation.

Technical Challenges and Future Outlook

Certain technical challenges still need to be addressed for wide-scale use and deployment of wearables as part of the digital health era. One such challenge is their privacy and security. Wearables are physically small and can store a large amount of data; therefore, it is highly likely that the device and data stored may be lost or even hacked. Development of more secure and encrypted wearables, potentially with tracking capabilities, is desired to reduce privacy and security risks.

Another technical challenge is the personal calibration of wearable devices. Every person is uniquely different, and various factors affect personal health (e.g., family medical history, genetics, and diet); therefore, the symptoms for early disease diagnosis may differ for every person. Thus, personal calibration of devices and machine learning–based analysis of data that are tailored for individuals are required for more accurate and relevant monitoring of the health status of the patient using wearables. Big data approaches driven by large populations (e.g., specific to different countries, races, etc.) are extremely powerful but can be misleading for individuals, and therefore wearable technologies should be aware of "small data" created by an individual and be ready to accurately harness and interpret them. Otherwise, big data streams will shadow outliers and statistically create catastrophic events for some individual patients.

A further challenge is the misalignment of wearables, which affects their measurement quality and accuracy. A wearable device should be capable of operating and gathering accurate data even if there is a misalignment between the user and wearable. This demands smarter designs that can tolerate such misalignments, for example, by using computational approaches and guide star– like internal references or self-calibration protocols. This is especially important considering the variations in human physiology and the size or 3D conformation of different organs on which wearables operate.

The durability and robustness of wearables also require further improvement. Wearables should be capable of operating under different conditions, such as in humid or wet environments and warm temperatures. This would allow continuous parameter monitoring without losing performance during such activities as swimming, showering, or sunbathing (depending on the target application). Wearables made of smart textiles and stretchable electronics should be washable and dryable (including the batteries), and the electrodes should not break when bending or folding the device. Another wearable part that requires increased robustness is the battery; this is especially crucial in GPS tracking, which is an essential tool for wearables that consumes a considerable amount of battery power while in use. The development of new materials that have more energy density and efficiently convert solar, thermal, or mechanical power into electricity will aid in the extension of the battery life. Wireless charging of wearables may be another option. Finally, the next-generation wearables will be even smaller than current versions, and as the size decreases, the packaging of sensors should be carried out in an integrated manner so that they do not lose their efficiency in a small area and can offer slim and even lighter-weight wearable designs.

Industrial Applications of Sensors

The industry sector is an important emitter of greenhouse gas emissions. According to GeSi, it was responsible for 23% of total emissions in 2002 and used nearly half of all global electricity. Sensors and especially sensor networks are used in multiple ways in industrial applications. They enable real-time data sharing on industrial processes, on the "health state" of equipment and the control of operating resources to increase industrial efficiency, productivity and reduce energy usage and emissions.

As the variety of different sensor applications is immense across industry sectors, this topic describes three examples of industrial fields of application of sensors for: i) process control, ii) control of (physical) properties during the production process, and iii) equipment management and control.

In the field of process control, sensor and sensor networks deliver real-time data on the production process and are able to detect in situ variations in the process. Control can thus be moved from the finished product after the completed production run to the production process itself. Faults can be minimised reducing the percentage of deficient and reprocessed goods. Furthermore, a continuous monitoring of processes allows for efficient use of energy during production processes. An application example in the field of process control is an on line laser-ultrasonic thickness gauge which measures the thickness of steel tube walls under harsh conditions in mills. During production, it ensures that "tube walls are uniform and reduces the need to remove excess material from the walls of the tubes". As a consequence, product consistency can be improved and material saved while "reducing the time and energy used during production".

In the field of the control of physical properties during production processes, sensors and sensor networks measure different properties as well as the amount of available resources during production. This allows them to be employed in an efficient and thus precise manner resulting in energy savings and the reduction of pollutants. Examples are sensors measuring the temperature and composition of combustion gases and sensors measuring the concentration of hydrogen gas.

In the third field, equipment management and control, sensors monitor the "health of machines" as well as their usage. Sensors installed on different machines measure physical properties such as temperature, pressure, humidity or vibrations. The sensor nodes are able to communicate between each other and send data to the network where the data is processed. When critical values are achieved, the system immediately sends signals making predictive maintenance possible. This intelligent maintenance monitors the functionality of parts and ensures that they are replaced based on a degradation assessment rather than on replacement rules. Besides health monitoring, sensors also control motors during usage. Motors running at full capacity regardless of load can be inefficient and waste energy. Sensors allow the motor to adjust the power usage according to the required output. Wireless networks that link different sensors make machine-to-machine communication possible and have the potential to increase energy efficiency in whole factories.

As the examples have shown, many specific and niche sensor applications are used in factories. Consequently, interoperability of different systems becomes a crucial issue to connect different sensor systems and to maximise efficiency and energy savings. Some standards have already been launched such as the interface IEEE 1451 group of standards which aims at enabling plug-and-play of different sensors and sensor networks.

There is so far little information on the overall environmental impact of sensors and sensor networks across different fields of industrial applications. GeSi assessed the impact of one major industrial field of application: smart motor systems. In the following, the results of this analysis are discussed to give an example of the environmental impact of smart industrial applications.

According to GeSi motor systems account for 65% of total energy use by industry. Smart motors which adjust power consumption to outputs can have an important role in reducing this demand. The authors of the study estimate a worldwide abatement potential of 970 MtCO_2e in a BAU scenario. This is, on the one hand, due to an optimisation of motors' speed (abatement potential of 680 MtCO_2e) and, on the other hand, to ICT-driven automation (abatement potential of 290 Mt-CO_2e).

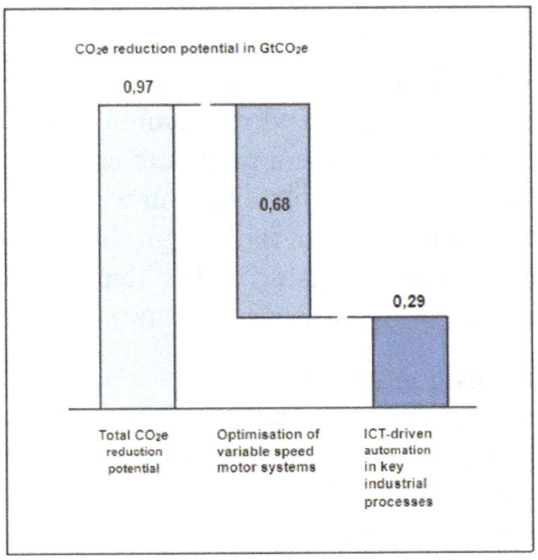

Positive Environmental Impact of Smart Motors

Table: Assumptions underlying the Calculations of Positive Impacts.

Lever	Assumptions for the calculations
Optimisation of variable speed of motor systems	30% increase in efficiency of industrial motor systems through optimisation 60% penetration of motor system optimisation technology.
ICT-driven automation in key industrial processes	15% decrease in total electricity consumption. 33% penetration of process optimisation technology.

The discussed example highlights that sensor technology has an important impact on energy and greenhouse gas emission savings for industrial automation and control. These savings can be especially high when different sensors and sensor networks communicate with each other. Besides the use of sensor technology, sound process planning, for example with process optimisation tools, also play an important role. Various initiatives have been created such as Motor Decision Matters, and Work Energy Smart which not only focus on technology but also on process planning and improvement.

Precision Agriculture and Animal Tracking

Sensors and sensor networks are important components of precision agriculture which aims at “maximum production efficiency with minimum environmental impact”. Land over-exploitation, one of the major concerns of intensive agriculture, leads to problems such as soil compaction, erosion, salinity and declining water quality. Sensors and sensor networks play a critical role in measuring and monitoring the health of the soil and water quality at various stages, from pre-to post-production. In the field of animal tracking, the movement of herds, the health of animals and the state of the pasture can be controlled via sensor networks. So far a number of sensor network systems have been developed and trials and field experiments are under way. However, concrete applications are at an early stage.

In precision agriculture, sensor networks can be used for: i) plant/crop monitoring, ii) soil monitoring, iii) Climate Monitoring and iv) Insect-Disease-Weed Monitoring.

In the field of plant/crop monitoring, wireless sensors have been developed to gather, for example, data on leaf temperature, chlorophyll content and plant water status. Based on these data, farmers are able to detect problems at an early stage and implement real-time solutions. The health and moisture of soil is a basic prerequisite for efficient plant and crop cultivation. Sensors contribute to real-time monitoring of variables such as soil fertility, soil water availability and soil compaction. Further, sensor nodes which communicate with radio or mobile network weather stations provide climate and micro-climate data. Sensors registering the temperature and relative humidity can contribute to detect conditions under which disease infestation is likely to occur.

The health of pastures can also be evaluated through high-resolution remote sensing tools. Healthy pastures usually “have a consistent cover of evenly dispersed perennial vegetation”. Remotely sensed satellite maps depict the location of persistent vegetation cover. Based on this information and information on the three dimensional shape of the landscape, Australian scientists calculate leakiness values and their changes over time. As a result, conditions of pastures can be measured and problematic areas detected.

Wireless sensors are further used for precision irrigation, and systems developed for remotely controlled, automatic irrigation. Sensors assume, for example, the tasks of irrigation control and irrigation scheduling using sensed data together with additional information, e.g. weather data. Finally, sensors are used to assist in precision fertilisation. Based on sensor data, decision support systems calculate the “optimal quantity and spread pattern for a fertilizer”.

Wireless sensor networks also contribute to a better understanding of the behaviour of cattle, such as their grazing habits, herd behaviour and the interaction with the surrounding environment. The information provided by these sensors helps famers to understand the state of the pasture and to find optimal ways to use these resources. To test sensor applications for cattle management, Wark et al. attached sensor nodes to cattle collars. Sensors communicated in a peer-to-peer fashion. Cattle collars pinged each other "with each ping containing an animal's GPS position and time of each ping transmission". Based on the positioning data of each node and inertial information, the cattle's individual and herd behaviour could be modelled and more general models could be developed. As a result, farmers are able to optimally manage environmental resources and plan grazing areas to prevent environmental problems such as overgrazing and land erosion. Current work focuses on the integration of sensor networks and radio frequency technology (RFID) as a significant number of cattle are equipped with RFID tags to record their ID as well as information such as cattle characteristics and food information.

The Environmental Impact of Precision Agriculture and Animal Tracking

Through the monitoring of the soil, climate and plants, a precise irrigation rate can be determined which may lead to a reduced consumption of water. Usually, fields are irrigated with uniform amounts of water. However, the variability in a field requires different amounts for different areas due to the combination of different crops and soil types. Various projects have been conducted to measure the extent of water savings. Damas et al. for example, tested an automated irrigation system for a 1500 ha area in Spain, with water savings of 30-60%. According to the USDA, another study found water savings of 5.7 million gallons on 279 acres in 2002. One study by King et al. showed no significant water savings for a variable rate irrigation system. However, the spatial variability in available water holding capacity (AWHC) of the soil was considered the main determinant influencing crop yield and the basis for a site-specific irrigation management (SSIM) system. The authors acknowledge that the "results from this study and others collectively suggest that AWHC may not be the best or only parameter to consider in delineating irrigation management zones. A systems approach to SSIM will likely be required that takes into account all known factors affecting yield ". Overall, the majority of studies showed a reduced consumption of water, but that deployment should be based on a thorough analysis of the area being irrigated and a comprehensive consideration of different factors which affect site-specific irrigation. Sensors and sensor networks can significantly contribute to this analysis by providing the required data.

Further important benefits of precision agriculture are reductions of fertilisers and pesticides. Both fertiliser and pesticide applications affect surface and groundwater quality, the quality of crops, soil properties and non-target species. Through monitoring the soil, the micro-climate and crops, it is possible to apply only the fertilisers and the pesticides crops need. Rates can be varied in real-time within fields based on different field and plant properties. Additionally, applications can be more precisely controlled in environmentally sensitive areas. Finally, a more targeted application of pesticides can reduce problems of pesticides resistances.

In the field of animal tracking, farmers are able to manage grazing areas based on information on herd behaviour. As a consequence, overgrazing of pastures as well as land erosion can be avoided. Limited pasture resources can thus be effectively managed.

Overall, sensors and sensor networks significantly contribute to a more sustainable use of natural resources. However, development of sensors and sensor networks for precision agriculture is in an early stage and sensor applications tend to be expensive. To date, farmers only take economic benefits into consideration when deciding on whether they should rely on precision agriculture. Governments can help farmers to recognise the environmental dimension by pointing out the economic benefits of improved soil and pasture quality as well as reduced applications of fertilisers and pesticides. Further, precision agriculture can be encouraged through technical assistance and conservation programmes.

Optical Chemical Sensors in Industries

Chemical sensors generally may function in four possible modes of exposure to the analyte:

- Remote mode
- Immersion
- Adsorption
- Absorption

Remote mode is used when the sensor is removed far from the analyte-containing medium and the sensor has either to act actively and use electromagnetic beam to probe the medium at a distance or act passive and detect the emissions from the tested medium. An example of the remote mode of operation would be spectral measurement, used even to assess the chemical properties of deep space objects. This mode of operation that generally belongs to remote sensing is sometimes excluded from the definitions of the chemical sensor because of the lack of physical contact with analyte.

Immersion is the situation when the sensor is wholly or partially surrounded by the medium, but no further binding processes occur at its surface or within its bulk. This would be the case when a sensing probe is dipped into fluid and the optical properties of fluid are directly measured. Alternatively, the measured fluid may fill a large cavity within the sensing probe. In both cases the bulk properties of the measured fluid are those that trigger the sensor response.

Adsorption is a process in which an atom or molecule of a liquid or gaseous substance is bound to a solid surface. The adsorbed layer may be extremely thin, even monatomic or monomolecular, and it may not cover the whole surface, and still be sufficient to trigger a response of the chemical sensor. The bound substance is called adsorbate, and the solid surface to which the adsorbate adheres is called adsorbent. Adsorption occurs due to the presence of surface forces. The result of adsorption is an increase of the concentration of adsorbate at the adsorbent surface. A typical example of adsorption is condensation at the surface. Adsorbate may be physisorbed (bound by van der Waals forces or by electrostatic forces) or chemisorbed (bound by covalent bonding). The process opposite to adsorption is desorption – the removal of adsorbate particles from the adsorbent.

If the binding of substance occurs within the volume of a solid object, the effect is called absorption, the substance dissolved within the volume is the absorbate and the volume permeated by the absorbate is the absorbent. Absorption is not a consequence of surface forces. A typical example of absorption is incorporation of hydrogen atoms within palladium crystal lattice.

If the binding of particles from liquid or gas occurs within pores, like in nanoporous materials, the process will occur within the volume of the object and still it will be denoted as adsorption, because binding is a consequence of surface forces.

An example of such volume adsorption is binding of gases in zeolites or in activated carbon.

Adsorption and absorption both belong to the group of processes called sorption. Chemical sensors generally may any or both of the two sorption mechanisms for their function. In the case of optical devices, the increased concentration of the analyte (sorbate) changes the optical properties of sorbent material. These changes are detected by the readout beam.

The adsorption kinetics is usually described by isotherms that define the amount of the adsorbed substance at a constant temperature. For gas adsorption one typically uses the simple Langmuir isotherm that assumes a monolayer adsorption. If multilayers are formed by the adsorption of particles onto already adsorbed particles, one utilizes Brunauer-Emmett- Teller (BET) isotherm. An illustration can be seen in figure. There are other isotherms that are used to analyze adsorption at the surface of chemical sensors.

More than one process may occur concurrently within the same structure. For instance, an analyte may adsorb to the surface of a structure and at the same time diffuse and permeate its bulk.

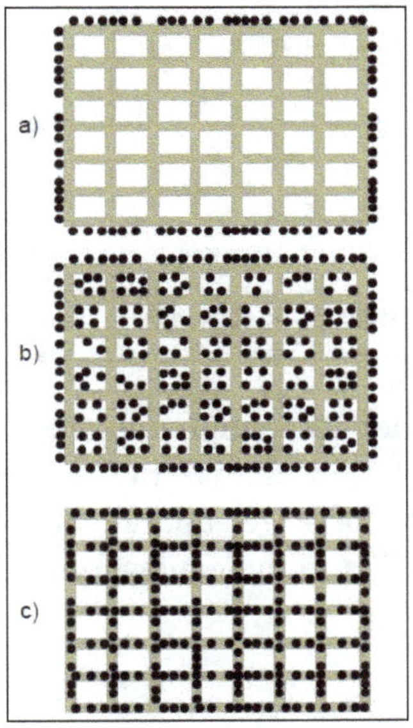

Sensor exposure modes: a) surface adsorption; b) bulk adsorption in nanopores; c) absorption (permeation/dissolution). In all three cases the active part of the sensor is a porous ordered structure.

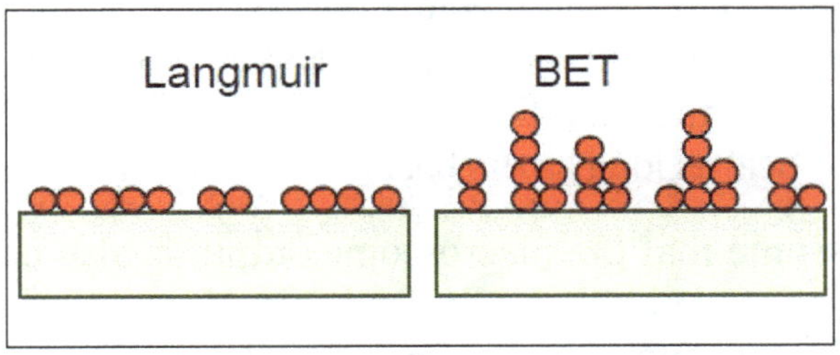

Examples of Langmuir (left) and BET (right) adsorption.

Affinity Sensors

In our further consideration we analyze adsorption-based sensors. The medium around a chemical sensor typically contains several competing species that may simultaneously adsorb on the same substrate. It may happen that two or more of these adsorbates cause the same modulation of sensor parameters and thus an identical output. This makes a great problem with sensor sensitivity. A way to avoid it is to utilize affinity sensors, i.e. the devices that specifically bind to a target analyte.

The term "affinity sensor" was used by Schultz et al in 1970 ties to describe a glucose sensing device. In affinity sensors one utilizes material that preferentially binds the targeted analyte. This material is denoted as receptor and can be deposited to the active surface of the sensor to promote target-specific binding. This material is sometimes also denoted as ligand. An illustration of this case can be seen in figure.

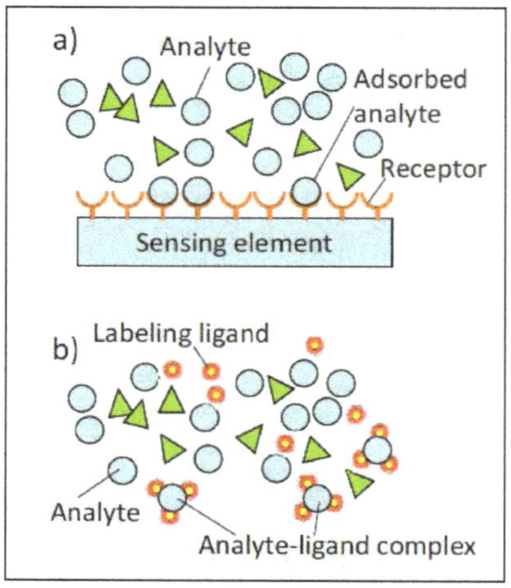

Two mechanisms of operation of affinity-based sensors. a) receptor layer at the sensor surface binds targeted analyte and discards undesired similar species; b) ligand (e.g. fluorescent molecule) labels the analyte particles.

One also utilizes the term ligand to denote signal-triggering substance (for instance, fluorescent labeling molecule) that binds to the target analyte in order to ensure its detection. Thus affinity sensor may be a device utilizing a layer of target-analyte binding receptor material deposited onto its surface, but also a device utilizing ligand molecules for labeling.

Macroscopic Readout

The external world can only see changes of the macroscopic properties of a chemical sensor caused by the presence of an analyte. These may be any electric, thermal, electric, chemical, etc. properties observable by our instruments. One could imagine devices that change temperature due to some kind of interaction between the active part and the analyte, modulate electric potential, change optical transmission, even, in some more exotic situations, change shape (like in bimaterial devices). The sensor may be called optical as long as the basic interaction that causes modulation is optical.

In a strict sense we can assume that the macroscopic output is also optical. This is the simplest situation, and one can always assume that a device converting optical interaction to non-optical signal basically consists of two or more units, one generating optical output, the other converting the optical output to non-optical. Thus for the sake of simplicity, in this text we consider only the optical output.

Two most obvious optical parameters to modulate are the scattering parameters of the sensor – its optical transmittance T and optical reflectance R. These are connected with the optical absorptance as $A = 1 - T - R$. Of course, the interaction can also cause optical emission or modify the already existing emission. One can also utilize devices that change the polarization of the utilized optical waves (chirality or handedness of the light). Thus the following macroscopic optical outputs can be utilized for sensor readout:

1. Reflection
2. Transmission
3. Absorption
4. Emission
5. Polarization
6. Phase change

One should mention that the use of the above may not be obvious at a first glance. For instance, in conventional surface plasmon polariton-based devices one varies the measurement spatial angle in order to obtain the sensor. Yet, the underlying principle of signal readout is the determination of reflectance minimum related with plasmon resonance.

The following parameters may be varied when observing the quoted outputs:

1. Intensity (amplitude)
2. Wavelength (spectral dependence)
3. Temporal dependence
4. Spatial distribution

Different outputs and varying parameters can be combined when obtaining the macroscopic output from the sensor. For instance, when measuring reflection, one can observe its intensity at a given moment and for a given wavelength, or alternatively observe its distribution across wavelengths (reflection spectroscopy), temporal dependence (changes of reflection coefficient over time), spatial dependence (reflection may vary across the analyzed surface). This does not relate with the intrinsic transduction mechanism used to obtain a particular change.

Optical Transduction Phenomena

The optical phenomena that are being modulated are the heart of an optical chemical sensor. They are the transduction mechanisms itself that define the sensor behavior. These mechanisms further modulate transmission, reflection, absorption or emission of optical radiation.

The IUPAC definition of the main modes of transduction of optical sensors include absorbance, reflectance, fluorescence, refractive index, optothermal effect and light scattering. However, such a definition simultaneously uses several criteria instead of a single one, since for instance absorbance modulation actually represents a consequence of refractive index change (being proportional to the imaginary part of the complex refractive index that describes losses), reflectance changes are the consequence of the changes of the real part of the refractive index, etc. On the other hand, it does not take into account some other phenomena, like e.g. photoacoustic, electrochromic, etc. In this topic we opted to utilize instead different microscopic mechanisms that modulate the properties listed as macroscopic outputs. Some of them include:

1. Wave superposition
    - Interferometry
    - Optical resonators
    - Diffractive methods
2. Evanescent wave mechanisms
    - Total internal reflection
    - Surface plasmon polaritons
    - Extraordinary optical transmission
    - Surface enhanced Raman scattering (SERS)
    - Surface enhanced infrared absorption (SEIRA)
    - Dyakonov waves
3. Photoluminescence phenomena
    - Fluorescence
    - Fosforescence
4. Polarimetry/ellipsometry
    - Optical rotation (optical activity)
    - Optical rotatory dispersion
    - Circular dichroism
5. Scattering
    - Rayleigh scattering
    - Mie scattering
    - Tyndall scattering
    - Brillouin scattering
    - Raman scattering
    - Fano resonance

6. Nonlinear optical phenomena
    - Harmonic generation by frequency mixing
    - Optical rectification
    - Optical parametric generation, amplification, oscillation
    - Spontaneous parametric down conversion
    - Light-plasma interaction
    - Stimulated Brillouin scattering
    - Cross-polarized wave generation
    - Optical soliton interactions
    - Kerr effect
    - Four-wave mixing
7. Quantum optical phenomena
    - Upconversion
    - Electromagnetically induced transparency
    - Optically active quantum structures (e.g. quantum dots)
8. Photothermal effects
9. Photoacoustic and acousto-optic effects
10. Photoelasticity
11. Electro-optical mechanisms
    - Electroabsorption
    - Stark effect
    - Franz-Keldysh effect
    - Electrochromic effect
    - Pockels effect
    - Electron-refractive effect
    - Electro-gyration
12. Transformation optics
    - Superabsorption and "optical black holes"

- Superlensing
- Hyperlensing
- Sensor cloaking

Many of the above described sensing schemes reduce to refractometric ones since they are based on modification of the complex refractive index. Also, two or more may be combined in the same sensor.

Device Structures

We assume that an optical chemical sensor is basically an optical waveguide in a most general sense, i.e. a structure that supports electromagnetic (EM) waves. These waves may be propagating or localized, sinusoidal or exponentially decaying (evanescent); they may be transmitted, reflected, absorbed or emitted. The structure itself may have an arbitrary geometry and its electromagnetic properties can be described by a given spatial distribution of its complex refractive index:

$$n(\omega,\vec{r})=n_{\text{Re}}(\omega,\vec{r})+i\cdot n_{\text{Im}}(\omega,\vec{r})$$

where ω is the angular frequency, r is the spatial position vector, "Re" stands for real and "Im" for imaginary part of the refractive index. Thus in a general case n is frequency dispersive, inhomogeneous (spatially varying), anisotropic (with properties in a given point varying in dependence on the EM wave direction) and lossy.

A division of waveguiding structures for optical sensors can be done based on the dimensions of their characteristic features compared to the operating wavelength λ:

1. Superwavelength (characteristic dimensions much larger than λ)
   - Spectroscopic cells
2. Mesoscopic (characteristic dimensions comparable with λ)
   - Planar waveguides
   - Fiber optics
   - Photonic crystals
3. Subwavelength (characteristic dimensions much smaller than λ)
   - Surface plasmon polariton devices
   - Nanoplasmonics
   - Metamaterials

Planar Waveguides

Many chemical and biological sensors base their operation on all-dielectric optical waveguides that utilize total internal reflection (TIR) for the wave propagation. In order to ensure waveguiding, the dimensions of such an optical guide along at least one axis should be comparable to the operating wavelength. A generic structure of a chemical sensor based on planar optical waveguide is shown in figure.

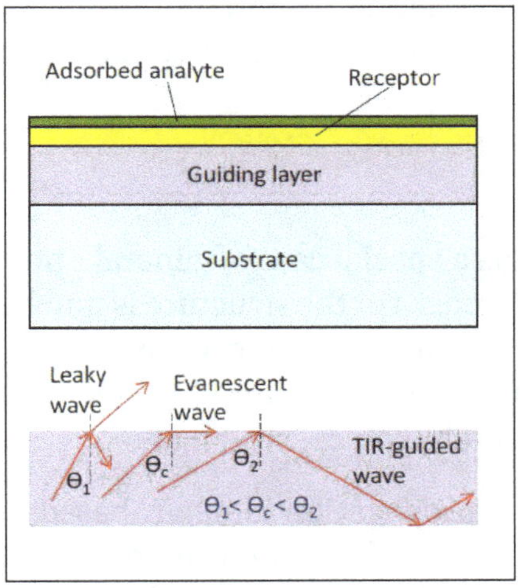

Generic structure of a chemical sensor based on optical waveguide (top) and the scheme of optical wave propagation in the guiding layer (bottom).

The bottom part of figure illustrates how TIR supports a propagating wave within the waveguide. If the refractive index within the guide is larger than that of its environment, the propagation of a light beam within it will depend on its angle of incidence on the boundary between the guide and the ambient. If this angle is small enough, a part of the beam will reflect from the boundary, and a part will refract. When increasing the angle, the output beam will refract at a increasingly larger angle. For the incidence under the critical angle qc no light will leave the guide, and no light will be reflected too. All of it will instead propagate as an evanescent wave strictly along the boundary. For even larger angles of incidence the beam will be fully reflected from the boundary, it will travel to the boundary at the opposite side of the guide, bounce from it, and thus propagate through the guide after repeated reflections from its boundaries - this is the TIR effect. Mode will be changed and the ensuing change in propagation will be detected by some of the previously quoted external means. Thus the waveguide-based sensors are typical refractometric sensors. However, any of the previously listed transducing mechanism could be in principle used for their operation. Also, besides the described affinity method, some other mechanism could be used as well, for instance bulk absorption that causes physical swelling and the resulting change of the waveguide dimensions.

Some types of rectangular channel waveguides. Top left: ridge waveguide; top right: rib waveguide; bottom left: strip-loaded waveguide; bottom right: embedded waveguide. Waveguiding layer is denoted by darker color.

Figure illustrates some typical planar waveguide structures utilized in chemical sensing. These are rectangular waveguides denoted as channel waveguides. Top left picture represents a ridge waveguide; top right is a rib waveguide. Bottom left is a strip-loaded waveguide where a lower index material is deposited over a strip waveguide. Finally, bottom right picture represents an embedded waveguide. Besides these geometries, different other refractive index profiles and geometries can be also used.

Optical Fibers

An optical fiber may be regarded as a special case of general optical waveguide where full cylindrical symmetry exists and where the length of the structure is much larger than its radial dimensions. Optical fibers are widely used in optical communications, but one of their important applications is in fiber optical sensing. They are used for sensing of a wide range of physical quantities, and one of their applications is chemical and biological sensing.

In conventional optical fibers the propagation is again based on the principle of total internal reflection, the same as in planar waveguides. A typical optical fiber consists of a core with a higher refractive index and a cladding with a lower refractive index. If the refractive index abruptly changes between the core and the cladding, such structure is denoted as step-index fiber. If there is a gradient of refractive index, it is denoted as graded index structure. Figure shows some typical situations. The case (a) represents a multimode fiber with step index profile. The core has a larger refractive index and thus a TIR-guided propagation is enabled, the same as in the case of planar guides. The case (b) represents a single-mode fiber (smaller diameter core), (c) is a fiber with a gradient of refractive index and (d) is the situation with multiple steps.

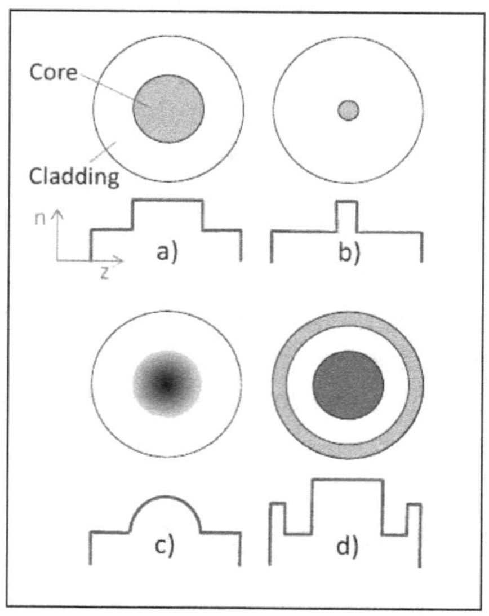

Some profiles of optical fibers: cross section and refractive index values.
Darker shading corresponds to higher refractive indices. a) multimode fiber
with step profile; b) single-mode fiber, step profile; c) graded index fiber; multiple-step fiber.

The usual way to utilize an optical fiber is to remove a part of cladding and let the evanescent field at the core boundary interact directly with the analyte. Another way is to utilize inverted profile fibers, for instance inverted graded index (IGI), where larger value of index are near the fiber surface.

As far as the readout is concerned, the most often used mechanisms are absorbance (as defined by Beer-Lambert-Bouguer extinction law), fluorescence measurement, diffraction on grating inscribed directly in the fiber and surface plasmon polaritons.

One can use one fiber for illumination and another for detection, or alternatively the same fiber is used for both.

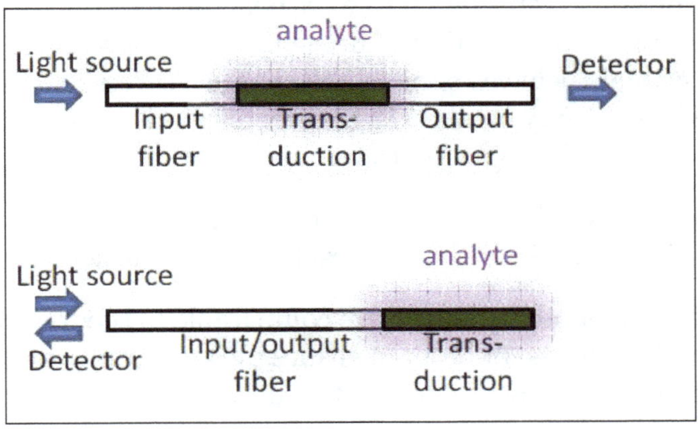

Readout schemes for chemical fiber optic sensors. Top: separate input and output (two-fiber scheme); bottom: single fiber scheme for simultaneous input and output.

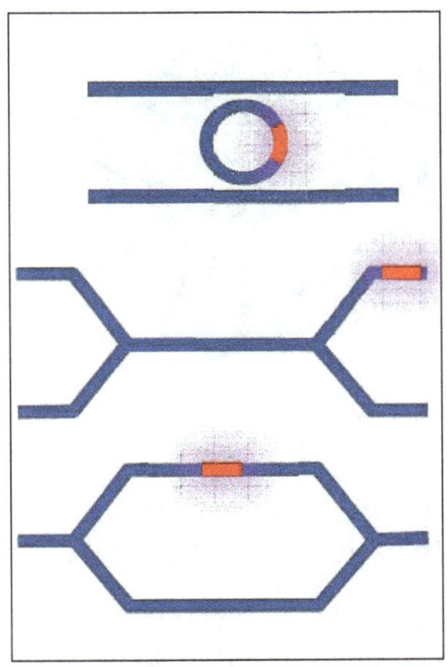

Top: Fiber ring resonator for chemical sensing; middle: Fiber-based Mach-Zehnder interferometer as sensor; bottom: Michelson interferometer for chemical sensing.

One can fabricate resonators from fibers by performing coupling between a straight fiber section and a fiber ring, or between two fibers divided by a ring, as illustrated in figure on top. This ensures the appearance of sharp peaks in spectral characteristics, while the index change introduced by analyte results in a shift of the center frequency of the resonator.

Another possibility is to fabricate different interferometers using two branches of fibers, one of which is then exposed to analyte. Examples of such fiber structures for chemical sensing are Mach-Zehnder and Michelson interferometers.

Photonic Crystals

Photonic crystals are artificial structures with periodically changing refractive index in 1D, 2D or 3D, where the contrast between the low and high values of refractive index is sufficient to ensure the existence of the wavelength range in which no electromagnetic waves can propagate through the structure. The range of forbidden wavelengths is the photonic bandgap (PBG) and in that range a photonic crystal behaves as a near-perfect mirror. It could be said that photonic crystals represent a generalization of Bragg interference mirrors and also of Bragg gratings. Light in photonic crystals behaves similarly to charge carriers in semiconductors – besides the appearance of a bandgap, energy bands exist that correspond to the valence and the conductive band, it is possible to introduce defect and surface modes, etc.

The behavior of a photonic crystal is defined by the dimension of their unit cell and by the value of the refractive index (or dielectric permittivity) of each of its constituent parts.

The features of a unit cell of a photonic crystal are comparable to the wavelength and their behavior is based on the diffraction effects. In order to ensure the appearance of a full photonic bandgap, a photonic crystal structure must have a sufficiently large contrast of refractive index between its "high" and "low" parts.

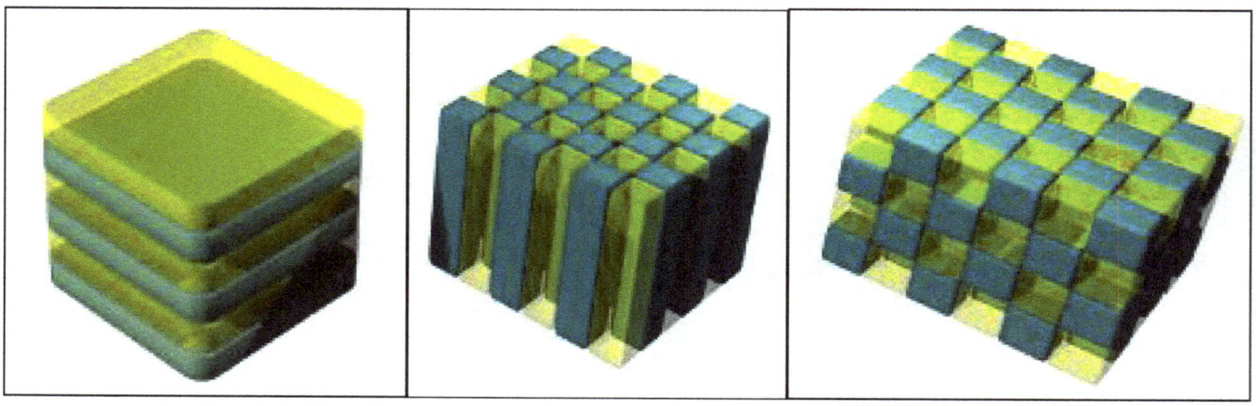

1D, 2D and 3D Cartesian geometries (from top to bottom) for PBG-based chemical sensors.

Photonic crystals were introduced in 1987. Until now they found a plethora of applications, among which an important place belongs to their use for chemical and biological sensing.

Structurally, there are three types of photonic crystals for chemical sensors, high-contrast Bragg layers, basically functioning as interference-based dielectric mirrors (1D structures); planar waveguides or resonators utilizing 2D geometry and full 3D crystal.

The properties of photonic crystals can be tuned by the presence of analyte by two mechanisms:

1. Changing refractive index due to the presence of analyte (for instance by adsorption or absorption).
2. Changing the unit cell dimensions, the simplest case being the shrinkage or swelling of the structure due to the presence of analyte, for instance by utilizing hydrogel that interacts with the targeted substance.

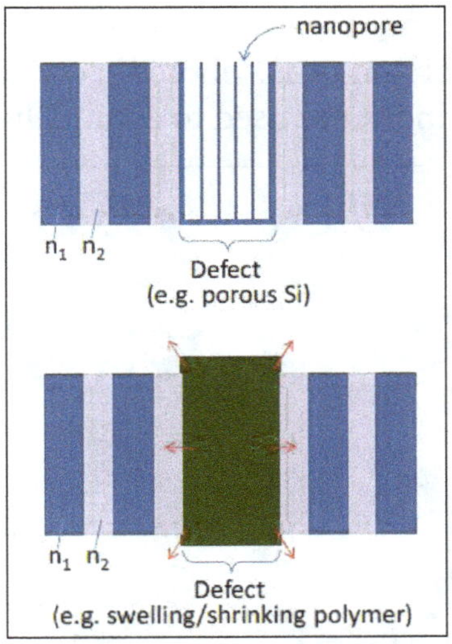

USe of 1D photonic crystals as chemical sensors. Top: absorption of analyte by porous defect (refractive index change); bottom: absorption of analyte in polymer that causes swelling (photonic crystal dimensions change).

The two basic methods of sensing of chemical substances are illustrated in figure for the case of 1D photonic crystal. A layer with different thickness and/or refractive index is introduced in the structure. Depending on its properties, it functions as an acceptor or donor defect, as per equivalence with semiconductors. In the transmission spectrum it shows as a sharp peak (a dip in reflection). In the case shown in top the defect is made of a porous material with subwavelength holes, for instance nanoporous silicon. Upon exposure to chemicals, analyte fills the pores and modifies the effective value of refractive index. Instead of the scheme shown in figure, one can alternatively use structures where each layer is porous, but with different densities in different parts (higher hole-to-solid ratio means a lower effective refractive index).

The scheme shown in bottom illustrates the possibility to change the dimensions of the photonic crystal. To this purpose, a polymer layer is introduced as a defect. The exposure to analyte initiates specific binding between the polymer part and the analyte and the result is swelling or shrinking of the defect, which causes a shift of the defect peak. Again, as in the previous case, one can utilize the variable volume material not only for the defect, but for the whole photonic crystal or only for its high or low refractive index parts. In this case the whole photonic crystal is infused with analyte and the resulting thickness change modifies the spectral characteristics.

Obviously, the same kind of tunability is applicable in the case of 2D and 3D photonic crystals as well.

A special kind of PBG structures often used for chemical sensing are the photonic crystal fibers, also called the microstructured optical fibers. These are a special kind of 2D photonic crystals that are produced in the form of fibers. A high refractive index contrast in these structures is obtained by introducing hollow (air filled) channels within the fiber, or using some other type of microstructuring to ensure the contrast. There are several basic types of photonic crystal fibers, as shown in figure. The first type are the so-called holey fibers, structures with a hollow core and hollow channels around it (top left). The core may also be solid (top right). There are combination of hollow channels with conventional step index optical fibers (bottom left) where the channels ensure better index contrast and thus better guiding. Finally, there is a Bragg-type fiber (bottom right) which basically represents a cylindrical version of 1D photonic crystal and where concentric layer with alternating low and high refractive index are used to confine light.

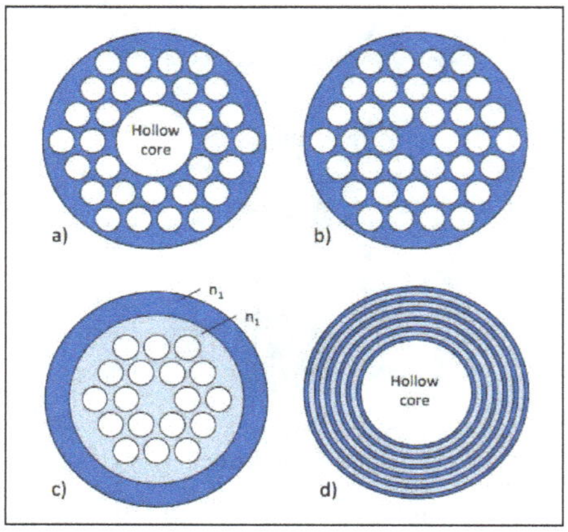

Different types of photonic crystal fibers for chemical sensors. a) hollow core fiber (holey fiber); b) solid-core fiber; c) mixed-type (holeassisted fiber); d) Bragg fiber ("OmniGuide").

Of the listed structures, those with hollow core are the most convenient for sensing, since the core can be filled with analyte, the interrogating light is simultaneously confined within the same space and the optical paths can be very long. However, structures with solid core were also proposed for chemical sensing.

Surface Plasmon Polariton Devices

Surface plasmons polaritons (SPP) may be defined as oscillations of free charge carriers in conductive material near an interface with dielectric which are coupled with electromagnetic radiation propagating across the interface. The relative dielectric permittivity e in the conductive part is negative in the range of interest, and positive in the dielectric part. Typically the conductive material is metal (usually gold or silver. Other alternative materials can be sometimes used, for instance transparent conductive oxides (indium tin oxide, doped zinc oxide, tin oxide, etc.), semiconductor materials, intermetallics, graphene, etc. These conductors are all denoted as plasmonic materials.

The electromagnetic wave in SPP is confined to the interface and evanescent (exponentially decaying) in the perpendicular direction. In chemical sensors one also encounters the spatially confined SPP that exist on a conductive nanoparticle (localized surface plasmons polaritons, LSPP). The science dedicated to the research and application of SPP phenomena and devices is called plasmonics.

In the simplest case, SPPs at the interface between two semi-infinite media with opposite signs of dielectric permittivity are not plane electromagnetic waves. Their energy is concentrated in the narrow region near the boundary plane. SPPs are TM (transverse-magnetic) polarized, and because of that they are called polaritons. In other words, magnetic field and wavevector of the SPP lay in the plane of interface, while electric field of the wave is both perpendicular and parallel to the wavevector components. TE polarized component of electromagnetic field cannot form a surface wave at a simple metal-dielectric interface.

A SPP wave has its largest value in the interface between the ε-positive and ε-negative part and decays exponentially in the both normal directions. If an analyte adsorbs to the sensor surface in an ultrathin layer, this will mean that the regions where the SPP is concentrated and the region where the analyte is present will overlap. Thus an SPP-based sensor will be able to “see” even minute amounts of analyte. Actually the conventional SPP sensors have long been able to distinguish even a few percent change of the composition of a monatomic or monomolecular layer.

The wavevector of a SPP is much larger than that of a propagating optical beam in the free space for the same frequency. Thus it is necessary to match the wavevector between the two. The most often used approach for the conventional SPP chemical sensors is the Kretchmann method. The interrogating beam arrives to the metalized flat surface at exactly the critical angle and thus the created evanescent wave is matched to the SPP.

At the surface plasmon resonance there will be a strong dip in the reflection spectrum that corresponds to a given angle of incidence and output. When an ultrathin layer of analyte adsorbs to the surface, its refractive index changes the propagating conditions of the evanescent SPP waves and thus the angle is changed. The angular shift is detected in the readout signal, and its amount is determined by the refractive index of analyte. Thus a SPP sensor is actually a refractometric sensor, basing its operation on the refractive index change within the near-surface layer with a subwavelength thickness.

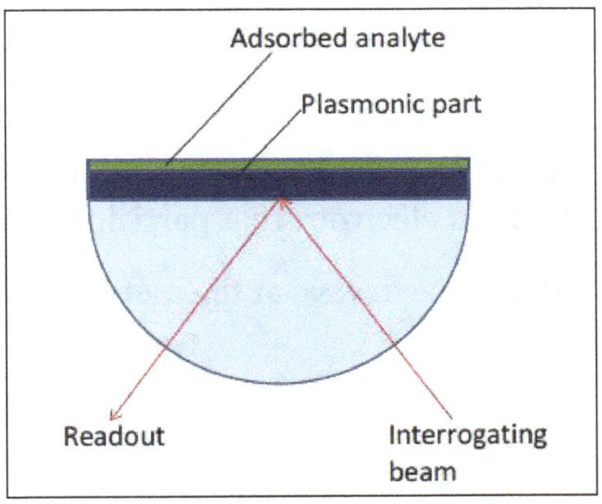

Kretchmann configuration for SPP sensor readout.

The resonant nature of the SPPs actually represents an advantage for sensing purposes, since it causes steep gradients in their frequency dispersion, which results in large changes of scattering parameters for very small changes of the surface refractive index. Absorption losses in SPP structures are very large due to collisions of free carriers in the ε-negative material, which leads to shorter propagation paths, but this is not an issue for chemical sensing purposes.

The relative dielectric permittivity e of plasmonic materials is negative below plasma frequency, and its dispersion is well-described by the electron resonance model of Drude, also denoted as Drude-Sommerfeld model:

$$\varepsilon = \varepsilon_\infty - \frac{\omega_p^2}{\omega(\omega + i\Gamma)},$$

where ω_p is the plasma frequency, Γ is the damping factor (i.e. the imaginary part of the complex dielectric permittivity), and ε_∞ is the asymptotic dielectric permittivity. The plasma frequency is:

$$\omega_p^2 = \frac{n_e e^2}{m^* \varepsilon_0},$$

where n_e is the free electron concentration in epsilon-negative part, e is the free electron charge ($1.6 \cdot 10^{-19}$ C), ε_0 is the vacuum permittivity ($8.854.10^{-12}$ F/m), and m^* is the electron effective mass.

The damping factor is:

$$\Gamma = \frac{e}{\mu m^*}$$

where μ is the mobility of free electrons.

In the case of interband transitions between the valence and the conduction bands, dielectric permittivity is given by Lorentz model:

$$\varepsilon = \varepsilon_\infty + \frac{\omega'^2_p}{\left(\omega_0^2 - \omega^2\right) - i\Gamma'\omega},$$

where ω_o is the resonant frequency of the electron oscillator, while the apostrophe denotes values related with the concentration of bound electrons that participate in the interband transitions.

The propagation constant of a TM polarized wave at the metal (m)-dielectric (d) interface is given as:

$$\beta = \frac{\omega}{c}\sqrt{\frac{\varepsilon_m \varepsilon_d}{\varepsilon_m + \varepsilon_d}},$$

Thus a change of the dielectric permittivity (and therefore the refractive index) of the dielectric part at the interface will modify the propagation constant, thus changing the propagation of the SPP. This is the basis of the use of SPP structures for sensing.

The applicability of SPP resonance for chemical sensing enhancement has been recognized very early, and indeed some of the first proposed applications of the SPP were in refractometric sensing in thin layers at the ultrathin boundary layer between metal and dielectric. Currently this is probably the most often used optical method for chemical sensing.

Optical Metamaterial Sensors

The possibility to structure material at nanometer level has brought to a new paradigm, the electromagnetic metamaterials. These can be defined as artificial structures with subwavelength features and with effective optical properties that exceed those of natural media. Theoretically envisioned in 1967, practically described in 1999 and experimentally proven in 2001, metamaterials showed themselves applicable in a vast number of areas. The first described metamaterials were those with negative refractive index, but subsequently the term has spread to other structures with unusual values of refractive index, for instance zero refractive index.

Since one is able to tailor a metamaterial structure almost at will, this means that its dispersion relation could be fully tailored. It is thus possible to design modes with superluminal group velocities ("fast light"), near-zero ("slow light") and, as mentioned above, negative ("left-handed light," propagating in the direction opposite to that of the phase velocity).

The possibility to design a frequency dispersion at will ensures tailoring of spatially strongly localized near-field modes (due to shortening of the wavelength for a given frequency), thus obtaining a very high density of states compared to the free space. The same energy that a propagating wave would carry and at the same frequency is here compacted into a much smaller space, thus ensuring much higher wave densities and much shorter wavelengths for that frequency. This ensures highly enhanced interaction of optical radiation with the surrounding material. This kind of engineering of optical interaction with matter ensures its maximization in the active area of a sensor, with a consequence that its response is not only vastly increased, but also localized to a specific nano-volume. Thus a chemical sensor based on optical metamaterials will offer a significantly improved sensitivity. At the same time, a plethora of novel modes of operation will exist. Contrary to e.g. SPP sensors, no special readout schemes like Kretschmann prisms are necessary and the sensor designer freedom is much larger.

Probably the best known negative index metamaterial for the optical range is the fishnet, the structure that consists of a metal-dielectric-metal sandwich perforated by an array of nanoholes.

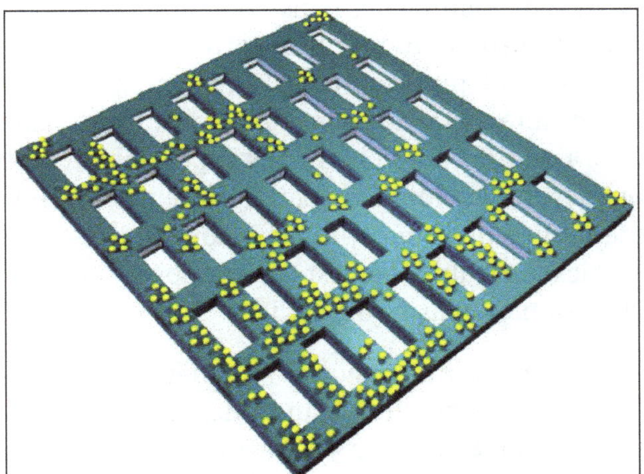

Fishnet metamaterial with rectangular holes. Spheres denote analyte atoms or molecules.

The whole structure is dipped into analyte-containing fluid, i.e. the absorption scheme is used to obtain the response shown in figure. This causes the shift of the dip in spectral reflection. This is a setup obtained by applying microfludics where analyte flows through nanoapertures, as described in for extraordinary optical transmission case. Thinner fishnets and smaller apertures are more strongly modified by analyte presence.

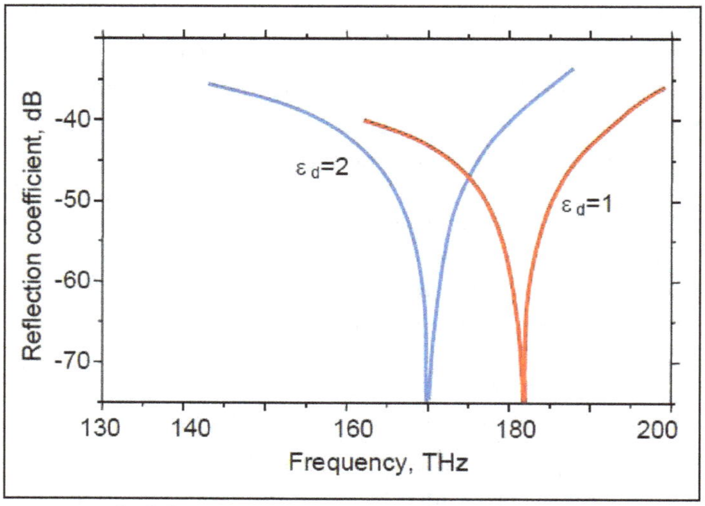

Spectral response of a fishnet metamaterial. Aperture size 320 nm x 280 nm, metal t hickness 20 nm, dielectric thickness 200 nm, TM polarized in-plane response observed.

Due to their wider range of available values of refractive index, metamaterial structures ensure much better control over electromagnetic fields. Thus a whole new field of electromagnetic optics emerged, the transformation optics. This ensured different applications, like hyperlenses for conversion between near-field and propagating far-field modes, cloaking devices, superconcentrators and superabsorbers. All of these structures can be utilized in chemical or biological sensing, ensuring a large number of novel modes of operation of these devices.

Nanoplasmonic Structures

Nanoplasmonic structures are nanocomposites consisting of metals and dielectrics – or, in a more general case, of materials with positive and with negative relative dielectric permittivity). They represent a generalization of surface plasmon polariton structures – instead of having a simple flat boundary between semi-infinite metal and semiinfinite dielectric, the boundary here may have virtually any shape. A plethora of different field modes is possible in such structures. They ensure full control over both frequency dispersion and spatial dispersion, resulting in very high field enhancements and unprecedented resolutions. The behaviour of these structures is covered by the emerging science of plasmonics. Probably the best contemporary sensors of chemical and biological analyte belong to the class of nanoplasmonic devices.

Nanoplasmonic structures can be one-dimensional – planar superlattices with alternating metal and dielectric layers, 2D – metallic nanowires, nanorods, etc. embedded in dielectric, like nanometer-sized analogue of artificial dielectrics, or 3D – arrays of metallic nanoparticles embedded in dielectric). If the structure of nanoplasmonic composites is ordered and periodic, they are called plasmonic crystals. Nanoplasmonic structures can be also quasiperiodic (plasmonic quasicrystals), aperiodic or even random.

The basic building block of a nanoplasmonic structure, i.e. the artificial "atom" of such a plasmonic metamaterial, can have any form. This is in contrast to the natural materials, where there is no choice regarding the shape of the building blocks. The shapes of this metallic "atom" within the dielectric host may vary from simple sphere to very complex form from regular to irregular.

A fact of large importance for sensing is the possibility to produce additional field localizations by changing the shape of the plasmonic inclusions and their geometry. Similar to the edge phenomena in macroscopic world, the plasmonic inclusions with sharp corners will localize electromagnetic fields, the amount of localization being larger for sharper corners. Also, plasmonic nanoparticles with small interparticle distance (compared to the skin depth in metal) will cause strong coupling of surface plasmons polaritons between these particles, creating large field localizations in the space between them. This will cause high nonlocal effects and extreme localizations of electromagnetic field. A plethora of new modes and novel effects of interest for chemical sensing appears in such structures.

One of the typical situations is encountered if randomly distributed plasmonic nanoparticles are utilized. The may be placed on a surface or mixed with a fluid that contains analyte.

The field intensity around the nanoparticles is high due to the establishment of localized surface plasmons polaritons (LSPP). When analyte is present, its refractive index will change the electromagnetic conditions and the result will be for instance color shift. Fluorescence schemes can be also used in this case.

Another simple structure used for sensing are plasmonic nanomembranes. These are freestanding structures with thickness below 100 nm and extremely high aspect ratios, making their lateral dimensions reach millimeter, even centimeter range. When a surface plasmon polariton travels along one side of such a structure, it couples electromagnetically with the SPP on the other side. The result is the establishment of a long range surface plasmon polariton, which decays more slowly away from the surface and propagates at much larger distances than the usual SPP. This means that its "information distance" from the surface at which it can sense analytes is much larger than in conventional SPP structures. Because of that it is able to detect much larger analytes, including macromolecules and even different living cells.

A nanoplasmonic structure of a large importance for chemical sensing are extraordinary transmission arrays. If one opens nanoholes with dimensions much smaller than the operating wavelength in an optically opaque metal layer, according to the classical Bethe theory no light should be transmitted through the nanoholes because of the polarization of light. However, in 1998 Ebessen experimentally proved that a near 100% transmission can be obtained through such aperture array and denoted the effect extraordinary optical transmission (EOT). It was proven later that the effect is a consequence of the establishment of surface plasmons polaritons at the front and the back surface and the coupling of the SPP between these surfaces.

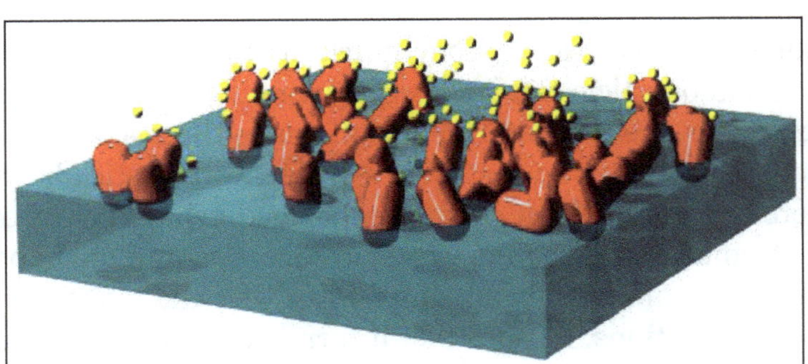

Nanoparticles placed on a substrate (capsule shapes) surrounded by analyte (small spheres).

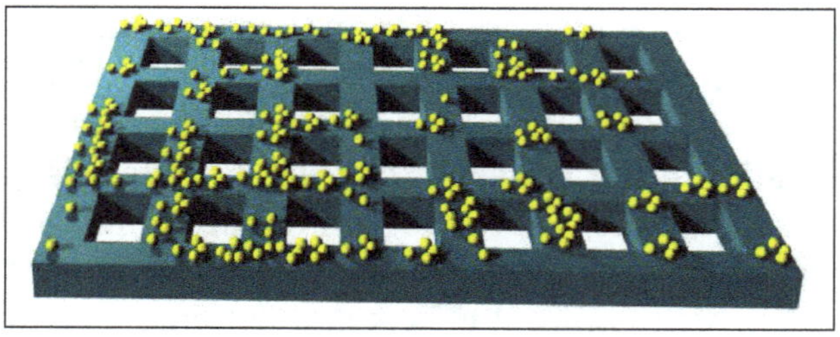

Extraordinary optical transmission structure.

The field intensity within the nanoapertures becomes several orders of magnitude larger than that impinging at the surface. Thus such structures ensure strong enhancement of nonlinear effects which is beneficial for selectivity of chemical sensing since different effects will occur in different analytes. At the same time, field concentration ensures high sensitivity chemical sensing at low analyte concentrations. Figure shows an EOT array.

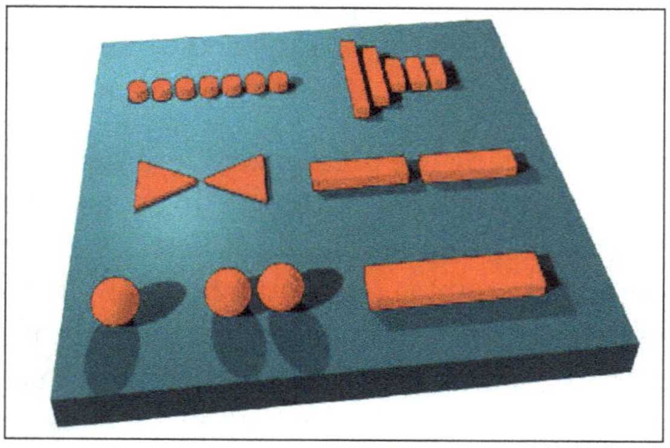

Various types of nanoantennas. Top left: linear nanoparticle array; top right: Udo-Yagi nanoantenna; middle left: bow tie antenna; middle right: two-wire nanoantenna. Bottom left: nanosphere, the simplest nanoantenna; bottom middle: nanodimer, two coupled spherical nanoantennas; bottom right: nanorod.

Similarly to fishnet metamaterials, EOT structures are very convenient for microfluidic chemical sensing schemes, where fluid that contains analyte is pumped through nanoapertures.

Probably the most sensitive nanoplasmonic sensors are those based on nanoantennas. Nanoantenna or optical antenna is a plasmonic device that converts propagating waves into evanescent ones and back with a high efficiency. Field localization depends on the nanoantenna design and can be made deeply subwavelength. Thus interaction with photodetector active region can be vastly enhanced.

Nanoantennas are isolated structures, i.e. they function without a connection to any circuitry. Thus a simple spherical nanoparticle can be basically regarded as a dipole nanoantenna, bottom left. Other nanoantenna types met in literature include nanodimer, consisting of two nanosphere at a small (subwavelength) distance so that they are electromagnetically coupled. The distance between the two spheres in a nanodimer is denoted as feed gap and electromagnetic field localizations in it can be extremely strong. Other types are nanorod antennas (bottom right), bow tie antennas (middle left) that combine very small feed gap with the effect of sharp edges and thus ensures very high field localizations. Besides that it furnishes a broader bandwidth of optical frequencies. Two nanorods coupled together make a two-wire antenna (middle right). An array of nanodisks makes a complex linear nanoparticle array antenna. Stacking nanorods can make an equivalent of the classical Udo-Yagi antenna (top right). A vast number of different other nanoantennas has been reported until now.

In 2012 a golden nanorod antenna has been used to ensure sensing of single non-absorbing molecules. The same type of nanoantenna was utilized for in-situ, real time interactions with biomolecules in infrared spectrum. Different types of plasmonic antennas are being considered for various applications in chemical and biological sensing.

Sensors Applications in Transport and Logistics

Information and communication technologies (ICTs) and sensor networks in particular have the potential to contribute to increased efficiency in both freight and passenger transport as well as a potential reduction of overall transportation. On the one hand, increased use of ICTs can avoid freight and passenger transport through a higher degree of virtualisation, digitisation and teleworking. Digital content is delivered electronically and virtual conferences and teleworking reduce passenger transport. On the other hand, increased use of ICTs can contribute to better management of transport routes and traffic, higher safety, time and cost savings as well as reductions of CO_2 emissions.

Sensors and sensor networks play a vital role in the increase of transport efficiency. For example, sensor technology contributes to better tracking of goods and vehicles which might result in lower level of inventories and thus energy savings from less inventory infrastructure as well as a reduced need for transportation. Furthermore, sensors and sensor networks are pivotal parts of many intelligent transportation systems (ITS).

An intelligent transportation system (ITS) can be defined as "the application of advanced and emerging technologies (computers, sensors, control, communications, and electronic devices) in transportation to save lives, time, money, energy and the environment". The ITS can be categorised into intelligent infrastructure and intelligent vehicles.

Many of these applications are based on sensors and sensor networks. In the field of intelligent infrastructure sensors in pavements are used for road traffic monitoring systems to measure the intensity and fluidity of traffic (vehicle count sensors) and to provide information for traffic lights which are then controlled. These sensors are further able to detect whether, for example, public buses are approaching so that the green phase of traffic lights can be extended, allowing buses to keep their schedules. They also transmit information to update public transport panels. New sensor applications include. In addition, sensors are used for motorway tolling purposes where they detect vehicle RFID tags and retrieve the required information. Sensors also monitor the state of physical infrastructures such as bridges by detecting "vibrations and displacements".

Table: Intelligent Transportation Systems.

Intelligent Infrastructure		
Arterial and Freeway Management	Crash Prevention and Safety	Traffic Incident Management
▪ Traffic Signal Control, Lane Management	▪ Warning Systems	▪ Surveillance, Detection

▪ Surveillance, Enforcement	▪ Pedestrian Safety	▪ Response, Clearance
Emergency Management	Electronic Payment and Pricing	Roadway Operations
▪ Hazardous Material Management	▪ Toll Collection	▪ Asset Management
▪ Emergency Medical Services	▪ Multi-Use Payment	▪ Work Zone Management
Transit Management	▪ Traveller Information	Road Weather Information
▪ Operations and Fleet Management	▪ Pre-trip and En-Route Information	▪ Surveillance and Prediction
▪ Transportation Demand Management	▪ Tourism and Events	▪ Traffic Control
Information Management	Commercial Vehicle Operations	Intermodal Freight
▪ Information Warehousing Services	▪ Carrier Operations, Fleet Management	▪ Freight and Asset Tracing
▪ Archived Data Management	▪ Credentials Administration	▪ International Border Crossing
Collision Avoidance	Driver Assistance	Collision Notification
▪ Obstacle Detection	▪ Navigation, Route Guidance	▪ Advanced Automated Collision Notification
▪ Collision-Avoidance Sensor Technologies	▪ On-Board Monitoring	▪ In-Vehicle Crash Sensors

Intelligent vehicles are equipped with sensors for multiple purposes. Examples for different kinds of vehicles include: i) Trains in metros, especially driverless systems, use sensors to control the velocity and location of trains as well as stops at metro stations. ii) Buses rely on door sensors to detect whether doors are open to better locate them. Further applications include environmental sensors on buses and tramways that detect weather conditions, analyse traffic conditions and give alerts via on-board mini-computers; for applications in cars iii), current research projects focus on vehicle-to-vehicle communication based on data gathered by sensors. These (environmental) sensors collect information, for instance, on the location of the car, the speed and road and weather conditions. As cars pass each other, they are able to exchange the summarised information. Based on a detailed description of the environment, cars obtain traffic information and drivers are able to plan their routes more efficiently. In addition, trajectories of vehicles can be predicted resulting in a sophisticated risk assessment and thus increased traffic safety as drivers can be warned of dangerous driving conditions. A further example of car sensors is a tyre pressure monitoring system which delivers real time tyre pressure information to the driver. Besides improving safety, the system helps to "reduce the amount of emissions released into the atmosphere".

Overall, ITS systems make public and private transport more efficient, and potentially cheaper which may increase transport volumes (rebound effect) and the environmental impact might thus be negative. Results of studies analysing the environmental impact of smart transportation are mixed due to this effect, in contrast to other fields of application. As for smart buildings, the GeSI study concentrates on the positive impacts whereas the IPTS study covers positive and negative effects, including rebound effects.

The GeSI study estimates an abatement potential of 1.52 $GtCO_2e$ worldwide in the field of smart transport. As for the field of smart buildings, levers that can be attributed to sensors and sensor networks and further levers constitute the overall abatement potential. Levers that can include a positive impact of sensors and sensor networks are marked in blue solid shading in Figure, other levers have diagonal lines. According to GeSi, the most important levers include the optimisation of logistic networks and optimised collection and delivery planning.

Positive Environmental Impact of Smart Logistics

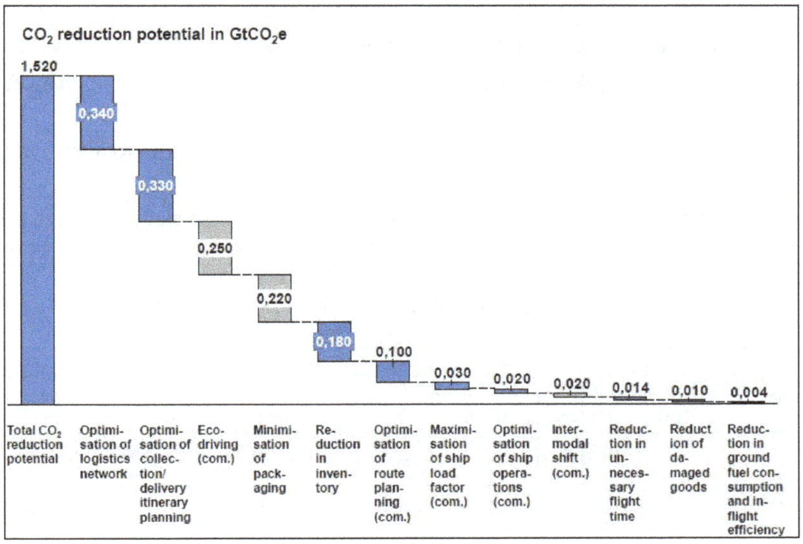

The Environmental Impact of Smart Transportation.

The IPTS study assesses the impact of ITS on passenger and freight transport. In contrast to the GeSI study, the authors find a significant increase of both passenger and freight transport volume across all scenarios with a low data uncertainty and a negative impact of ICTs in the field of ITS. This results in an increase of CO_2e emissions due to the use of ICT in transportation across all scenarios and best and worst case situations. Tableillustrates the results of the study. With the projected development of ICTs compared to a situation without this projected development, an increase in CO_2e emissions ranges from 1.9% in the best case and 2.7% in the worst case scenario.

Table: Assumptions underlying the Calculations of Positive Impacts.

Lever	Assumptions for the calculations
Optimisation of logistics network	14% reduction in road transport. 1% reduction in other modes of transport.
Intermodal shift	1% reduction in road transport owing to shift towards rail- and water-borne transport.
Reduction in inventory	24% reduction in inventory levels. 100% of warehouses and 25% of retail are assumed to be used for storage.
Optimisation of collection/delivery itinerary planning	14% reduction in road transportation.
Optimisation of truck route planning	5% reduction in carbon intensity of road transport owing to avoidance of congestion.
Eco-driving	12% reduction in carbon intensity owing to improved driving style.
In-flight fuel efficiency	1% reduction in fuel consumption achievable for 80% of t-km flown.

Reduction in ground-fuel consumption	32% reductions in ground fuel consumption achievable for 80% of flights. Impact calculated for average European fleet.
Reduction in unnecessary flight time (comm.)	1% reduction in fuel consumption achievable for 80% of t-km flown 32% reduction in ground fuel consumption achievable for 80% offlights.
Reduction in unnecessary flight time	3% reduction in flight time achievable for 80% of flights.
Maximisation of ship load factor	4% reduction in marine transport owing to improved utilisation of ships.
Optimisation of ship operations	3% increase in fuel efficiency, e.g. by adjusting ballasts and optimising speed.
Minimisation of packaging	5% reduction in packaging material, leading to a 5% reduction in all transports and in storage.
Reduction of damaged goods	0.2 % reduction in damaged goods achievable through better tracking (e.g. RFID) and conditions monitoring (e.g. bio-sensors).

Table: Impacts of Intelligent Transportation Systems (ITS) for different scenarios.

	Scenario A			Scenario B			Scenario C		
	Worst	Mean	Best	Worst	Mean	Best	Worst	Mean	Best
Freight transport tkm	13.3 %	13.4 %	13.5 %	27.3 %	27.8 %	28.2 %	12.4 %	12.5 %	12.6 %
Passenger transport pkm	5.5 %	5.3 %	5.2 %	6.1 %	6.1 %	6.1 %	5.6 %	5.7 %	5.7 %
Total energy consumption	1.9 %	2.1 %	1.9 %	2.6 %	2.8 %	2.5 %	1.9 %	2.0 %	1.9 %
Total GHG emissions	1.9 %	2.0 %	1.9 %	2.6 %	2.7 %	2.6 %	1.9 %	2.0 %	2.0 %
Scenario description									
Technology regulation	Incentives for innovation			Government intervention			Stakeholder approach		
Attitudes to ICT	Moderate, conservative			Receptive			Highly receptive		
ICT in business	High level of cooperation			High level of competition			Between A and B		
Attitudes to the environment	Moderate/controversial			High awareness and interest			High awareness and interest		

ITS render transport faster, more efficient and flexible and, as a consequence, cheaper, "leading to a full rebound effect". The demand for transportation increases and creates more transport leading to higher consumption of energy and to growing greenhouse gas emissions. According to the authors "higher transport efficiency is the key ICT effect increasing freight transport in 2020. This increase is in the range of 12% to 28%" for freight transport and of 5-7% for passenger transport. Although the increasing impact of ITS with the projected development of ICTs is significantly higher for scenario B, the absolute freight transport volume is the lowest one in this scenario as environmental costs are already internalised.

In the field of passenger transport, the increased time efficiency of passenger transport implies that a higher passenger transport volume can be attained in the same time which raises traffic performance. However, "ICT can slow the growth of private car passenger transport, avoiding 10-19% of future car traffic, despite the fact that it stimulates the growth of total passenger transport" due to better time utilisation. This time utilisation effect is supposed to increase the use of public transport in the modal split as ICTs can contribute to a more effective use of travel time to work.

The attractiveness of public transport can thus be increased and promote a shift from private cars to public transport. However, this effect also "relaxes the time budget and therefore enables more traffic".

In the field of freight transport, the full rebound effect which results from cheaper transport shows that freight transport is "highly sensitive to fuel prices". Raising fuel prices and thus internalising environmental costs could thus reduce demand significantly.

Overall, this overview of the GeSI and the IPTS studies shows mixed results of the impact of intelligent transport systems due to rebound effects. These effects also highlight that governments can have a crucial role in the field of smart transportation. As the IPTS study shows, increased efficiency of transportation should be paralleled with demand-side management. Internalisation of environmental effects by raising energy and fuel prices or the inclusion of transportation in emission trading could reduce demand for transport and thus reduce CO_2e emissions. Furthermore, governments can make use of ITS in public transport to render it more attractive and promote a modal shift from private cars to public transport. Measures not only include promoting a better working environment during travel through ICTs but also better services such as real-time time-table information and optimised route planning.

Remote Sensing

Remote sensing can be broadly defined as the collection and interpretation of information about an object, area, or event without being in physical contact with the object. Aircraft and satellites are the common platforms for remote sensing of the earth and its natural resources. Aerial photography in the visible portion of the electromagnetic wavelength was the original form of remote sensing but technological developments has enabled the acquisition of information at other wavelengths including near infrared, thermal infrared and microwave. Collection of information over a large numbers of wavelength bands is referred to as multispectral or hyperspectral data. The development and deployment of manned and unmanned satellites has enhanced the collection of remotely sensed data and offers an inexpensive way to obtain information over large areas. The capacity of remote sensing to identify and monitor land surfaces and environmental conditions has expanded greatly over the last few years and remotely sensed data will be an essential tool in natural resource management.

Electromagnetic Energy

The electromagnetic (EM) spectrum is the continuous range of electromagnetic radiation, extending from gamma rays (highest frequency & shortest wavelength) to radio waves (lowest frequency & longest wavelength) and including visible light.

The EM spectrum can be divided into seven different regions: gamma rays, X-rays, ultraviolet, visible light, infrared, microwaves and radio waves.

Remote sensing involves the measurement of energy in many parts of the electromagnetic (EM) spectrum. The major regions of interest in satellite sensing are visible light, reflected and emitted infrared, and the microwave regions. The measurement of this radiation takes place in what are known as spectral bands. A spectral band is defined as a discrete interval of the EM spectrum. For example the wavelength range of 0.4μm to 0.5μm (μm = micrometers or 10-6m) is one spectral band. Satellite sensors have been designed to measure responses within particular spectral bands to enable the discrimination of the major Earth surface materials. Scientists will choose a particular spectral band for data collection depending on what they wish to examine.

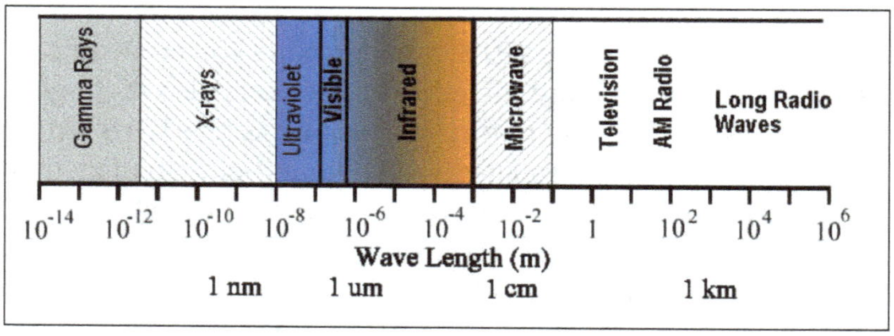

The design of satellite sensors is based on the absorption characteristics of Earth surface materials across all the measurable parts in the EM spectrum.

Reflection and Absorption

When radiation from the Sun reaches the surface of the Earth, some of the energy at specific wavelengths is absorbed and the rest of the energy is reflected by the surface material. The only two exceptions to this situation are if the surface of a body is a perfect reflector or a true black body. The occurrence of these surfaces in the natural world is very rare. In the visible region of the EM spectrum, the feature we describe as the color of the object is the visible light that is not absorbed by that object. In the case of a green leaf, for example, the blue and red wavelengths are absorbed by the leaf, while the green wavelength is reflected and detected by our eyes.

In remote sensing, a detector measures the electromagnetic (EM) radiation that is reflected back from the Earth's surface materials. These measurements can help to distinguish the type of land covering. Soil, water and vegetation have clearly different patterns of reflectance and absorption over different wavelengths.

The reflectance of radiation from one type of surface material, such as soil, varies over the range of wavelengths in the EM spectrum. This is known as the spectral signature of the material. All Earth surface features, including minerals, vegetation, dry soil, water, and snow, have unique spectral reflectance signatures.

Sensors and Platforms

A sensor is a device that measures and records electromagnetic energy. Sensors can be divided into two groups. Passive sensors depend on an external source of energy, usually the sun. The most common passive sensor is the photographic camera. Active sensors have their own source of energy, an example would be a radar gun. These sensors send out a signal and measure the amount reflected back. Active sensors are more controlled because they do not depend upon varying illumination conditions.

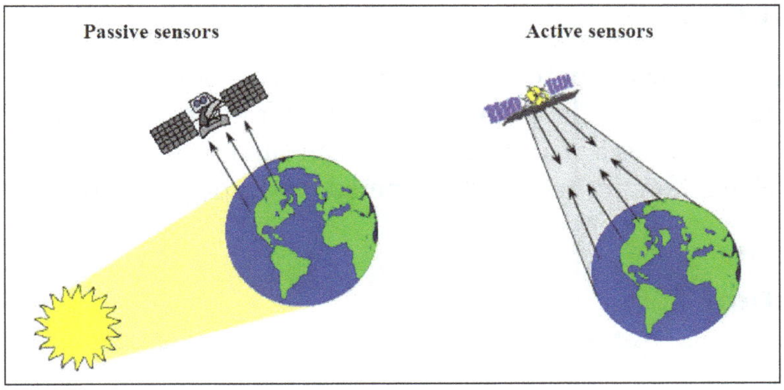

Orbits and Swaths

The path followed by a satellite is referred to as its orbit. Satellites which view the same portion of the earth's surface at all times have Geostationary orbits.Weather and communication satellites commonly have these types of orbits. Many satellites are designed to follow a north south orbit which, in conjunction with the earth's rotation (west-east), allows them to cover most of the earth's surface over a period of time. These are Near-polar orbits. Many of these satellites orbits are also Sun-synchronous such that they cover each area of the world at a constant local time of day. Near polar orbits also means that the satellite travels nortward on one side of the earth and the southward on the second half of its orbit. These are called Ascending and Descending passes. As a satellite revolves around the earth, the sensor sees a certain portion of the earth's surface. The area imaged is referred to as the Swath. The surface directly below the satellite is called the Nadir point. Steerable sensors on satellites can view an area (off nadir) before and after the orbits passes over a target.

Satellite Sensor Characteristics

The basic functions of most satellite sensors is to collect information about the reflected radiation along a pathway, also known as the field of view (FOV), as the satellite orbits the Earth. The smallest area of ground that is sampled is called the instantaneous field of view (IFOV). The IFOV is also described as the pixel size of the sensor. This sampling or measurement occurs in one or many spectral bands of the EM spectrum.

The data collected by each satellite sensor can be described in terms of spatial, spectral and temporal resolution.

Spatial Resolution

The spatial resolution (also known as ground resolution) is the ground area imaged for the instantaneous field of view (IFOV) of the sensing device. Spatial resolution may also be described as the ground surface area that forms one pixel in the satellite image. The IFOV or ground resolution of the Landsat Thematic Mapper (TM) sensor, for example, is 30 m. The ground resolution of weather satellite sensors is often larger than a square kilometer. There are satellites that collect data at less than one meter ground resolution but these are classified military satellites or very expensive commercial systems.

Temporal Resolution

Temporal resolution is a measure of the repeat cycle or frequency with which a sensor revisits the same part of the Earth's surface. The frequency will vary from several times per day, for a typical weather satellite, to 8—20 times a year for a moderate ground resolution satellite, such as Landsat TM. The frequency characteristics will be determined by the design of the satellite sensor and its orbit pattern.

Spectral Resolution

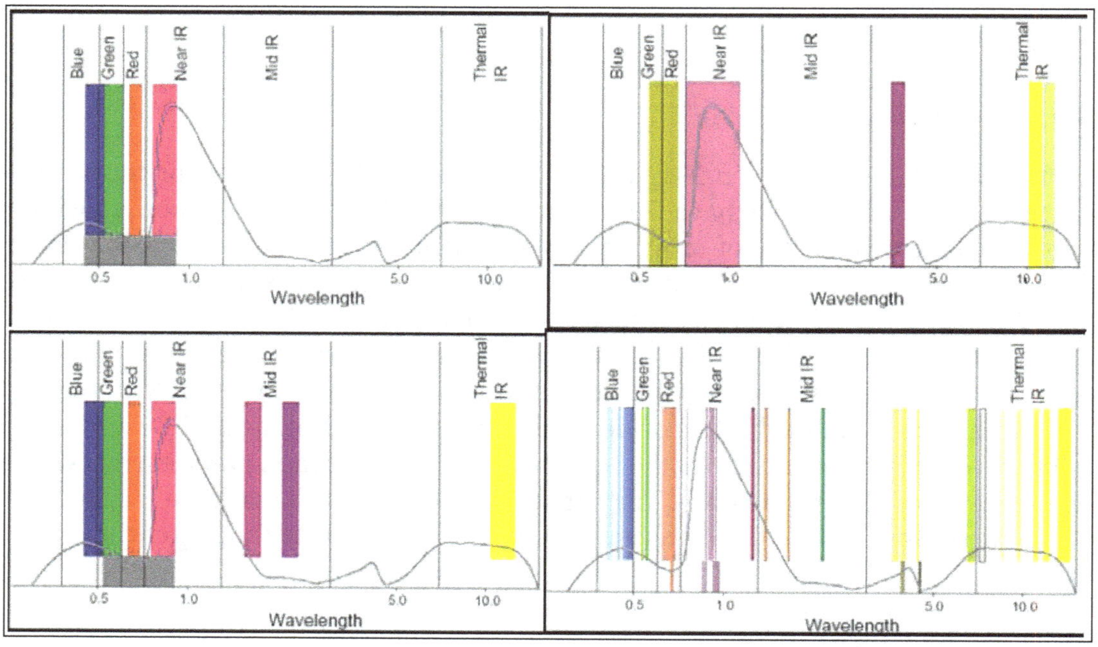

The spectral resolution of a sensor system is the number and width of spectral bands in the sensing device. The simplest form of spectral resolution is a sensor with one band only, which senses visible light. An image from this sensor would be similar in appearance to a black and white photograph from an aircraft. A sensor with three spectral bands in the visible region of the EM spectrum would collect similar information to that of the human vision system. The Landsat TM sensor has seven spectral bands located in the visible and near to mid infrared parts of the spectrum.

A panchromatic image consists of only one band. It is usually displayed as a grey scale image, i.e. the displayed brightness of a particular pixel is proportional to the pixel digital number which is related to the intensity of solar radiation reflected by the targets in the pixel and detected by the detector. Thus, a panchromatic image may be similarly interpreted as a black-and-white aerial photograph of the area, though at a lower resolution.

Multispectral and hyperspectral images consists of several bands of data. For visual display, each band of the image may be displayed one band at a time as a grey scale image, or in combination of three bands at a time as a color composite image. Interpretation of a multispectral color composite image will require the knowledge of the spectral reflectance signature of the targets in the scene.

Platforms

Aerial photography has been used in agricultural and natural resource management for many years. These photographs can be black and white, color, or color infrared. Depending on the camera, lens, and flying height these images can have a variety of scales. Photographs can be used to determine spatial arrangement of fields, irrigation ditches, roads, and other features or they can be used to view individual features within a field.

Infrared images can detect stress in crops before it is visible with the naked eye. Healthy canopies reflect strongly in the infrared spectral range, whereas plants that are stressed will reflect a dull color. These images can tell a farmer that there is a problem but does not tell him what is causing the problem. The stress might be from lack of water, insect damage, improper nutrition or soil problems, such as compaction, salinity or inefficient drainage. The farmer must assess the cause of the stress from other information. If the dull areas disappear on subsequent pictures, the stress could have been lack of water that was eased with irrigation. If the stress continues it could be a sign of insect infestation. The farmer still has to conduct in-field assessment to identify the causes of the problem. The development of cameras that measure reflectance in a wider range of wavelengths may lead to better quantify plant stress. The use of these multi-spectral cameras are increasing and will become an important tool in precision agriculture.

Satellite remote sensing is becoming more readily available for use in precision agriculture. The Landsat and the NOAA polar-orbiting satellites carry instruments that can be used to determine crop types and conditions, and to measure crop acreage. The Advanced Very High Resolution Radiometer (AVHRR) carried onboard NOAA polar orbiting satellites measure reflectance from the earth's surface in the visible, near infrared, and thermal infrared portions of the electromagnetic spectrum. This spectral sensitivity makes it suitable for measuring vegetative condition and because the satellite passes overhead twice a day, it can be used to detect rapidly changing conditions. Unfortunately, its use as a precision agriculture tool is limited because the spatial resolution of the sensor is nominally 1.1km. A possible application of this scanner would be to use the thermal infrared sensor to estimate daily maximum and minimum temperatures. These temperature estimates could then be used to determine degree-days that will drive pest development models. Degree-day models are an essential part of IPM programs and the enhanced spatial coverage provided by satellites would allow for assessment of spatial variability in predicted events that is not possible with data from sparsely spaced weather stations currently used for these models. Remotely sensed data can also be used to determine irrigation scheduling and adequacy of irrigation systems for uniformly wetting an entire field.

The sensors aboard the Landsat satellite measures reflected radiation in seven spectral bands from the visible through the thermal infrared. The sensors high spatial resolution (approximately 30m) makes it useful in precision agriculture. The spectral response and higher spatial resolution make it suitable for assessing vegetative condition for individual fields but the overpass frequency is only once every 16 days. The less frequent overpass makes it difficult to use these data for assessing rapidly changing events such as insect outbreaks or water stress. New satellites with enhanced capabilities are planned and remotely sensed data will become morewidely used in management support systems.

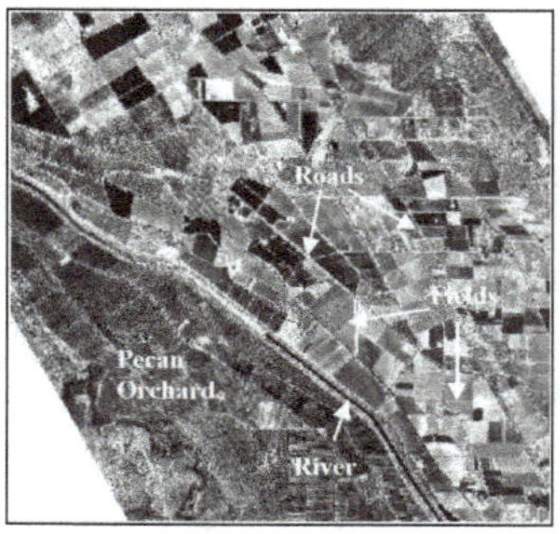

A black and white aerial photograph showing fields, roads, and irrigation ditches.

A high resolution aerial photograph showing individual trees within a orchard.

A color infrared photograph of a pecan orchard. Darker areas show stressed plants.

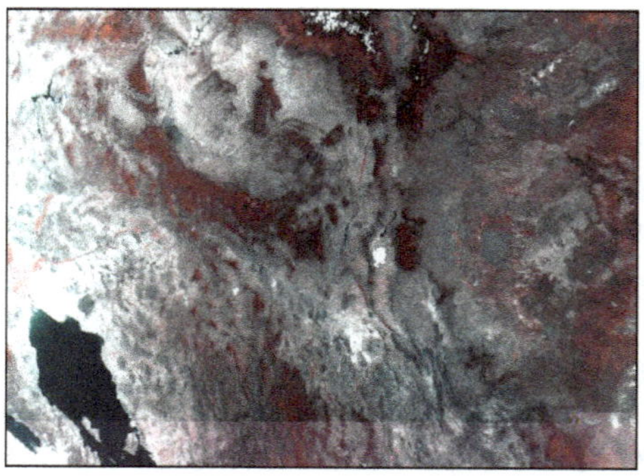

Advanced Very High Resolution Radiometer (AVHRR) image of the southwest Untied States. Image is centered on the Las Cruces.

A Landsat satellite image of farm land south of Las Cruces.

Common Satellites

Goes

5 spectral bands 1 - 41 km spatial resolution Geostationary.

Noaa Avhrr

5 spectral bands 1.1 km spatial resolution 1 day repeat cycle.

Landsat Tm

7 spectral bands 30m spatial resolution 16 day repeat cycle.

Modis

Multi- spectral bands 250-1000m spatial resolution (band dependent) 1day repeat cycle.

IKONOS

4 spectral Bands 4m spatial resolution 5 day repeat cycle.

Spectral Signatures of Natural and Human-made Materials

Remote sensing makes use of visible, near infrared and short-wave infrared sensors to form images of the earth's surface by detecting the solar radiation reflected from targets on the ground. Different materials reflect and absorb differently at different wavelengths. Thus, the targets can be differentiated by their spectral reflectance signatures in the remotely sensed images.

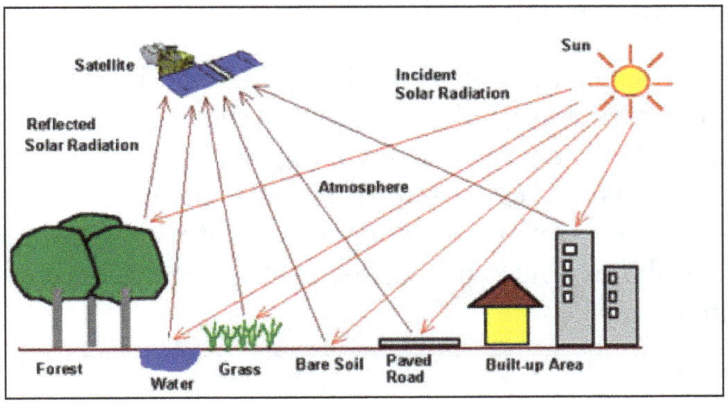

Spectral Reflectance Signature

When solar radiation hits a target surface, it may be transmitted, absorbed or reflected. Different materials reflect and absorb differently at different wavelengths. The reflectance spectrum of a material is a plot of the fraction of radiation reflected as a function of the incident wavelength and serves as a unique signature for the material. In principle, a material can be identified from its spectral reflectance signature if the sensing system has sufficient spectral resolution to distinguish its spectrum from those of other materials. This premise provides the basis for multispectral remote sensing.

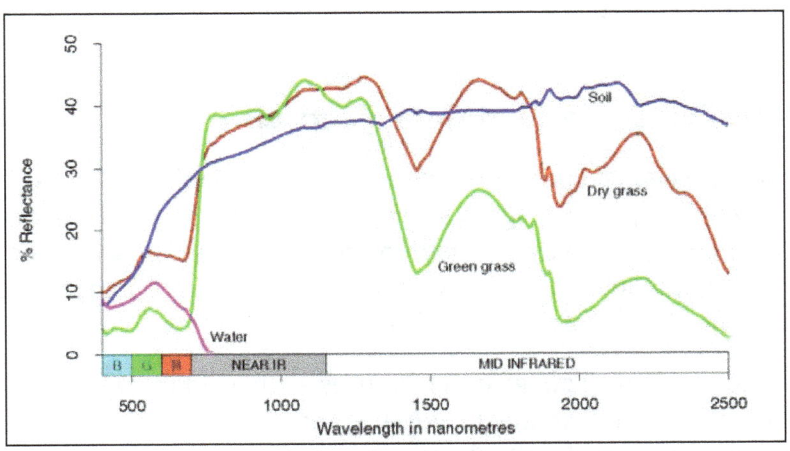

The reflectance of clear water is generally low. However, the reflectance is maximum at the blue end of the spectrum and decreases as wavelength increases. Hence, water appears dark-bluish to the visible eye. Turbid water has some sediment suspension that increases the reflectance in the red end of the spectrum and would be brownish in appearance. The reflectance of bare soil generally depends on its composition. In the example shown, the reflectance increases monotonically with increasing wavelength. Hence, it should appear yellowish-red to the eye.

Vegetation has a unique spectral signature that enables it to be distinguished readily from other types of land cover in an optical/near-infrared image. The reflectance is low in both the blue and red regions of the spectrum, due to absorption by chlorophyll for photosynthesis. It has a peak at the green region. In the near infrared (NIR) region, the reflectance is much higher than that in the visible band due to the cellular structure in the leaves. Hence, vegetation can be identified by the high NIR but generally low visible reflectance. This property has been used in early reconnaissance missions during war times for "camouflage detection".

The shape of the reflectance spectrum can be used for identification of vegetation type. For example, the reflectance spectra of dry grass and green grass in the previous figures can be distinguished although they exhibit the generally characteristics of high NIR but low visible reflectance. Dry grass has higher reflectance in the visible region but lower reflectance in the NIR region. For the same vegetation type, the reflectance spectrum also depends on other factors such as the leaf moisture content and health of the plants. These properties enable vegetation condition to be monitored using remotely sensed images.

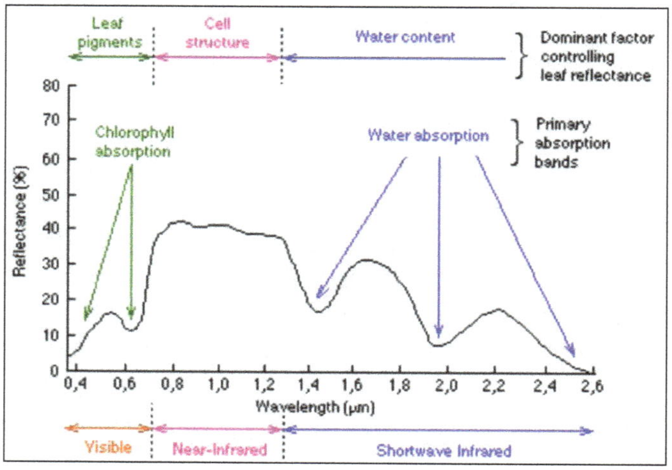

Geodesy, Geodetic Datums and Map Projections

Geodesy is the branch of science concerned with the determination of the size and shape of the Earth. Geodesy involves the processing of survey measurements on the curved surface of the Earth, as well as the analysis of gravity measurements. Knowing the exact location of a pixel on the Earth's surface (its spatial location) is an essential component of remote sensing. It requires a detailed knowledge of the size and the shape of the Earth.

The Earth is not a simple sphere. Topographic features such as mountain ranges and deep oceans disturb the surface of the Earth. The ideal reference model for the Earth's shape is one that can represent these irregularities and identify the position of features through a co-ordinate system. It should also be easy to use.

Flat Earth vs Curved Earth

The "flat Earth" model is not appropriate when mapping larger areas. It does not take into account the curvature of the Earth.

A "curved Earth" model more closely represents the shape of the Earth. A spheroid best represents the shape of the Earth because it is significantly wider at the equator than around the poles (Unlike a simple sphere). A spheroid, (also known as an ellipsoid) represents the equator as an elliptical shape, rather than a round circle. Surveying and navigation calculations can he performed over a large area when a spheroid is used as a curved Earth reference model.

Sea Level and the Composition of the Earth's Interior

The surface of the sea is not uniform. The Earth's gravitational field shapes it. The rocks that make up the Earth's interior vary in density and distribution, causing anomalies in the gravitational field. These, in turn, cause irregularities in the sea surface. A mathematical model of the sea surface can be formulated; however, it is very complex and not useful for finding geographic positions on a spheroid reference model.

Types of Geodetic Datum

Based on these ideas, models can be established from which spatial position can be calculated. These models are known as geodetic datums and are normally classified into two types geocentric datum and local geodetic datum.

A geocentric datum is one which best approximates the size and shape of the Earth as a whole. The center of its spheroid coincides with the Earth's center of mass. A geocentric datum does not seek to be a good approximation to any particular part of the Earth.

A local geodetic datum is used to approximate the size and shape of the Earth's sea surface in a smaller area.

Datums and GIS

Having a standard accurate datum set becomes increasingly important as multiple layers of information about the same area are collected and analyzed. The layers are developed into geographic information systems (GIS), which enable the relationships between layers of data to be examined. In order to function effectively, a GIS must possess one essential attribute. It must have the ability to geographically relate data within and across layers. For example, if a dataset about vegetation is being examined against the data sets for topography and soils, the accurate spatial compatibility of the two datasets is critical.

Map Projection Coordinates

A map projection is a systematic representation of all or part of of the Earth on a two-dimensional surface, such as a flat sheet of paper. During this process some distortion of distances, directions, scale, and area is inevitable. There are several different types of map projections. No projection is free from all distortions, but each minimizes distortions in some of the above properties, at the expense of leaving errors in others. For example, the commonly used Transverse Mercator projection represents direction accurately, but distorts distance and area, especially those farthest from the equator. Greenland, for example, appears to be much larger than it really is. The Transverse Mercator projection is useful for navigation charts.

Universal Transverse Mercator (UTM)

Universal Transverse Mercator (UTM) is a global spatial system based on the Transverse Mercator projection. UTM divides the Earth into 60 equal zones, each being 6 degrees wide. Each zone is bounded by lines of longitude extending from the North Pole to the South Pole. Imagine an orange consisting of 60 segments. Each segment would be equivalent to a UTM zone.

A rectangular grid coordinate system is used in most map projections. These coordinates are referred to as Eastings and Northings, being distances East and North of an origin. They are usually expressed in metres.

Under the UTM system, each East and North coordinate pair could refer to one of sixty points on Earth — one point in each of the sixty zones. Because of this, the zone number needs to be quoted to ensure the correct point on Earth is being identified.

Global Positioning System

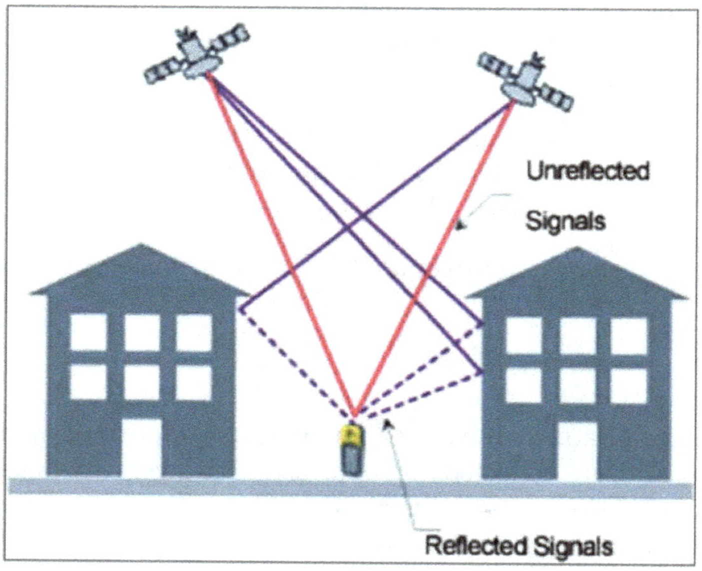

The Global Positioning System (GPS) is a satellite based system that gives real time three dimensional (3D) latitude, longitude, and height information at sub-meter accuracy. The system was developed by the United States military in the late 1970's to give troops accurate position and navigational information. A GPS receiver calculates its position on earth from radio signals broadcast by satellites orbiting the earth. There are currently twenty-four GPS satellites in this system. GPS equipment is capable of measuring a position to within centimeters but the accuracy suffers due to errors in the satellite signals. Errors in the signal can be caused by atmospheric interference, proximity of mountains, trees, or tall buildings. The government can also introduce errors in the signal for security purposes. This intentional degradation of the satellite signals is known as selective availability. The accuracy of the position information can be improved by using differential GPS. In differential GPS, one receiver is mounted in a stationary position, usually at the farm office, while the other is on the tractor or harvesting equipment. The stationary receiver calculates the error and transmits the necessary correction to the mobile receiver. GPS equipment suitable for precision agricultural cost several thousands dollars. Less expensive equipment is becoming available but the accuracy and capability is reduced.

Geographic Information System (GIS)

A Geographic Information System (GIS) is a computer-assisted system for handling spatial information. GIS software can be considered as a collection of software programs to acquire, store, analyze, and display information. The input data can be maps, charts, spreadsheets, or pictures. The GIS software can analyze these data using image processing and statistical procedures. Data can be grouped together and displayed as overlays. Overlays could be information such as soil type, topography, crop type, crop yield, pest levels, irrigation, and management information as shown.

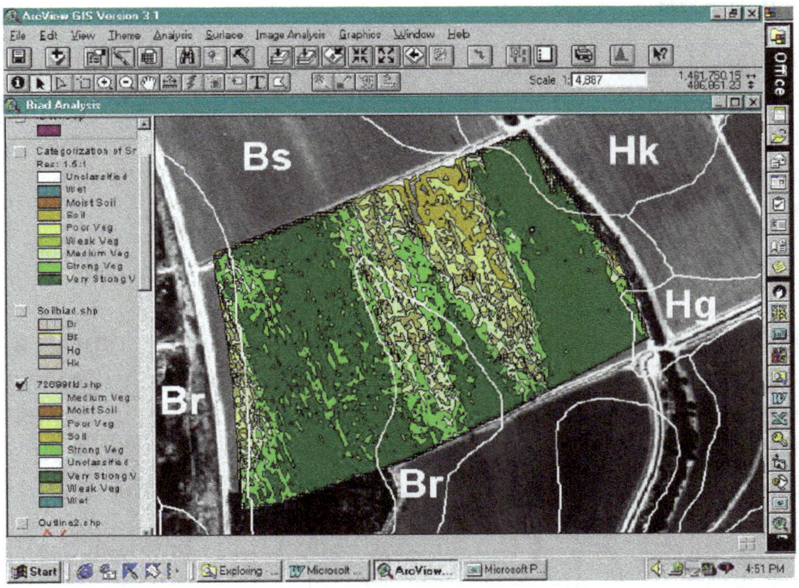

Relationships can be examined and new data sets produced by combining a number of overlays. These data sets can be combined with models and decision support systems to construct a powerful management tool. For example, we could assess how far a field was from roads or non-agricultural crops. This information could be important in pest infestation or in planning chemical application. We could also examine crop yield relationship to soil type or other factors as show in the following figure. A number of GIS software packages are now commercially available. Spatial data for the GIS is often collected using GPS equipment but another source of spatial information is aerial and satellite imagery.

Pixels, Images and Colors

Color Composite Images

In displaying a color composite image, three primary colors (red, green and blue) are used. When these three colors are combined in various proportions, they produce different colors in the visible spectrum. Associating each spectral band (not necessarily a visible band) to a separate primary color results in a color composite image.

Many colors can be formed by combining the three primary colors (Red, Green, Blue) in various proportions.

False Color Composite

The display color assignment for any band of a multispectral image can be done in an entirely arbitrary manner. In this case, the color of a target in the displayed image does not have any resemblance to its actual color. The resulting product is known as a false color composite image. There are many possible schemes of producing false color composite images. However, some scheme may be more suitable for detecting certain objects in the image.

Natural Color Composite

When displaying a natural color composite image, the spectral bands (some of which may not be in the visible region) are combined in such a way that the appearance of the displayed image resembles a visible color photograph, i.e. vegetation in green, water in blue, soil in brown or grey, etc. Many people refer to this composite as a "true color" composite. However, this term may be misleading since in many instances the colors are only simulated to look similar to the "true" colors of the targets.

For example, the bands 3 (red band), 2 (green band) and 1 (blue band) of a AVHRR image can be assigned respectively to the R, G, and B colors for display. In this way, the color of the resulting color composite image resembles closely what the human eyes would observe.

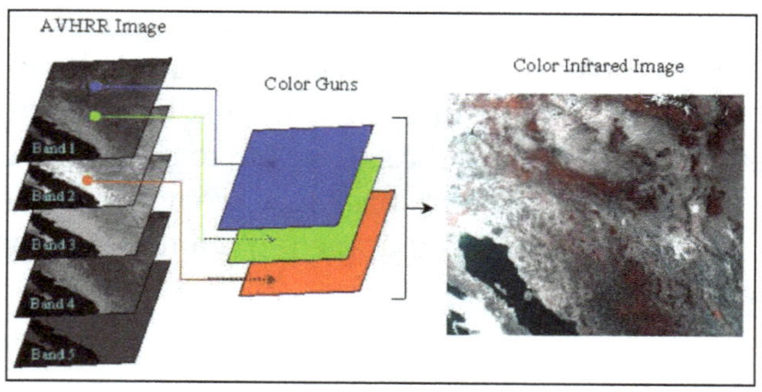

Image Processing and Analysis

Many image processing and analysis techniques have been developed to aid the interpretation of remote sensing images and to extract as much information as possible from the images. The choice of specific techniques or algorithms to use depends on the goals of each individual project. The key steps in processing remotely sensed data are Digitizing of Images, Image Calibration, Geo-Registration, and

Spectral Analysis. Prior to data analysis, initial processing on the raw data is usually carried out to correct for any distortion due to the characteristics of the imaging system and imaging conditions. Depending on the user's requirement, some standard correction procedures may be carried out by the ground station operators before the data is delivered to the end-user. These procedures include radiometric correction to correct for uneven sensor response over the whole image and geometric correction to correct for geometric distortion due to Earth's rotation and other imaging conditions (such as oblique viewing). The image may also be transformed to conform to a specific map projection system. Furthermore, if accurate geographical location of an area on the image needs to be known, ground control points (GCP's) are used to register the image to a precise map (geo-referencing).

Digitizing of Images

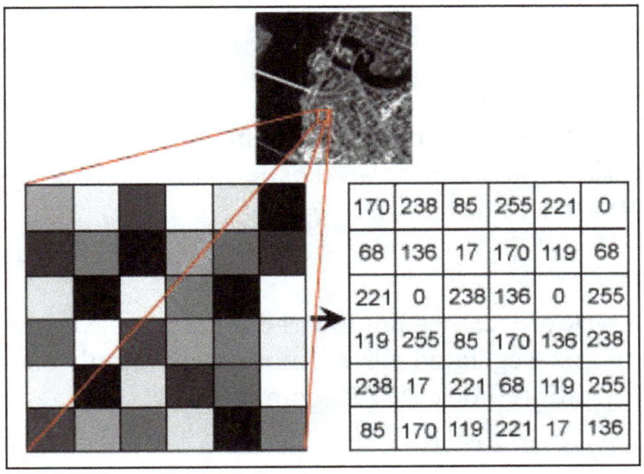

Image digitization is the conversion of an analogue image, such as a photograph, into a series of grid cells. The value of each cell is related to the brightness, color or reflectance at that point. A scanner is a simple way to digitize images. Many modern sensors now produce raw data in digital format.

Image Enhancement

In order to aid visual interpretation, visual appearance of the objects in the image can be improved by image enhancement techniques such as grey level stretching to improve the contrast and spatial filtering for enhancing the edges.

Image Classification

Different landcover types in an image can be discriminated using some image classification algorithms using spectral features, i.e. the brightness and "color" information contained in each pixel. The classification procedures can be "supervised" or "unsupervised".

In supervised classification, the spectral features of some areas of known landcover types are extracted from the image. These areas are known as the "training areas". Every pixel in the whole image is then classified as belonging to one of the classes depending on how close its spectral features are to the spectral features of the training areas.

In unsupervised classification, the computer program automatically groups the pixels in the image into separate clusters, depending on their spectral features. Each cluster will then be assigned a landcover type by the analyst.

Each class of landcover is referred to as a “theme” and the product of classification is known as a “thematic map”.

The information derived from remote sensing images are often combined with other auxiliary data to form the basis for a Geographic Information System (GIS). A GIS is a database of different layers, where each layer contains information about a specific aspect of the same area which is used for analysis by the resource scientists.

Image Interpretation

Vegetation Indices

Different bands of a multispectral image may be combined to accentuate the vegetated areas. One such combination is the ratio of the near-infrared band to the red band. This ratio is known as the Ratio Vegetation Index (RVI):

$$RVI = NIR/Red$$

Since vegetation has high NIR reflectance but low red reflectance, vegetated areas will have higher RVI values compared to non-vegetated aeras. Another commonly used vegetation index is the Normalised Difference Vegetation Index (NDVI) computed by:

$$NDVI = (NIR - Red)/(NIR + Red)$$

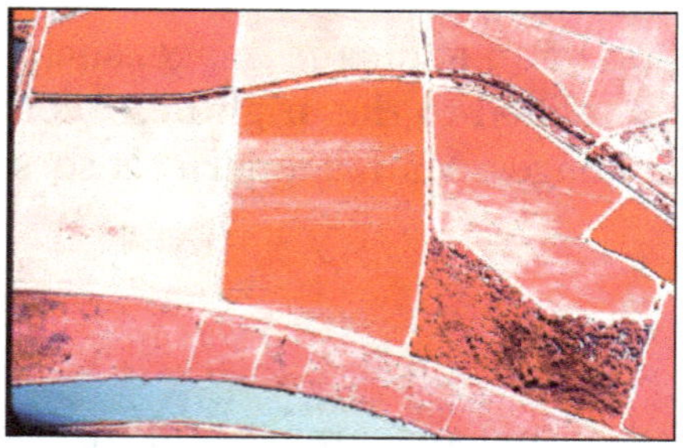

Vegetation maps are produced by generating a normalized difference vegetation index from a infrared image and then doing a vegetation classification.

Color infrared photographs collect information in the green, red and near infrared light reflectance spectrum. Green vegetation reflects very strongly in the near infrared light range and therefore infrared images can detect stress in many crops before it is visible with the naked eye.

The Normalized Difference Vegetation Index (NDVI) is used to separate green vegetation from the background soil brightness. It is the difference between the near infrared and red reflectance normalized over the sum of these bands.

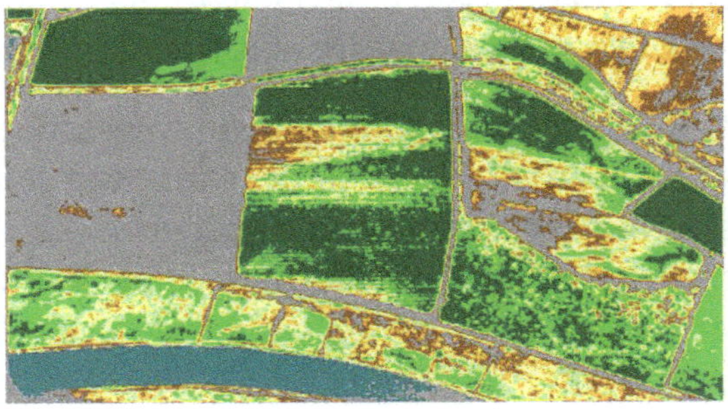

NDVI = (IR-Red)/(IR+Red), these NDVI maps can then be classified into vegetation categories and displayed as a vegetation maps with different colors representing different levels of vegetation. In the map on the left browns and yellow represent bare soil and shades of green represent vegetation, darker greens are stronger vegetation.

Current Applications of Remote Sensing

Forestry Applications

Satellite imagery is used to identify and map:

- The species of native and exotic forest trees.
- The effects of major diseases or adverse change in environmental conditions.
- The geographic extent of forests.

This application of satellite imagery has led to the extensive use of imagery by organizations that have an interest in a range of environmental management responsibilities at a state and national level.

Greenhouse Gases: Sinks and Sources

Forests are often referred to as carbon sinks. This description is used because during photosynthesis, carbon dioxide, the major greenhouse gas, is taken from the atmosphere and converted into plant matter and oxygen.

Climate change has serious implications for Australia and overseas countries alike. Sustainable land management is essential for effective greenhouse gas management; hence, it is important to acquire data on land cover in Australia. Remotely sensed land cover changes are used in calculations of our national emission levels, and data collected on a national scale will enable governments to develop responses to land clearing.

Vegetation Health

Vegetation can become stressed or less healthy because of a change in a range of environmental factors. These factors include lack of water, concentration of toxic elements/herbicides and infestation by insects/viruses. The spectral reflectance of vegetation changes according to the structure and health of a plant. In particular, the influence of chlorophyll in the leaf pigments controls the response of vegetation to radiation in the visible wavelength. As a plant becomes diseased, the cell structure of a plant alters and the spectral signature of a plant or plant community will change.

The maximum reflection of electromagnetic radiation from vegetation occurs in the near infrared wavelengths. Vegetation has characteristically high near-infrared reflectance and low red reflectance. Air-borne scanners using narrow spectral bands between 0.4 urn and 0.9 urn can indicate deteriorating plant health before a change in condition is visible in the plant itself.

Biodiversity

Vegetation type and extent derived from satellite imagery can be combined, with biological and topographic information to provide information about biodiversity. Typically, this analysis is done with a geographic information system.

Change Detection

Satellite imagery is not always able to provide exact details about the species or age of vegetation. However, the imagery provides a very good means of measuring significant change in vegetation cover, whether it is through clearing, wildfire damage or environmental stress. The most common form of environmental stress is water deficiency.

Geology

Remote sensing is useful for providing information relevant to the geosciences. For example, remote sensing data are used in:

- Mineral and petroleum exploration,
- Mapping geomorphology,
- Monitoring volcanoes.

Land Degradation

Imagery can be used to map areas of poor or no vegetation cover. A range of factors, including saline or sodic soils, and overgrazing, can cause degraded landscapes.

Oceanography

Remote sensing is applied to oceanography studies. Remote sensing is used, for example, to measure sea surface temperature and to monitor marine habitats.

Meteorology

Remote sensing is an effective method for mapping cloud type and extent, and cloud top temperature. In many of the applications identified above remotely sensed data are used with a range of other Earth science data to provide information about the natural environment. This analysis of Earth science data from a range of sources is usually done in a geographic information system (GIS).

Collocation

Collocation is a procedure used in remote sensing to match measurements from two or more different instruments. This is done for two main reasons: for validation purposes when comparing measurements of the same variable, and to relate measurements of two different variables either for performing retrievals or for prediction. In the second case the data is later fed into some type of statistical inverse method such as an artificial neural network, statistical classification algorithm, kernel estimator or a linear least squares. In principle, most collocation problems can be solved by a nearest neighbor search, but in practice there are many other considerations involved and the best method is highly specific to the particular matching of instruments. Here we deal with some of the most important considerations along with specific examples.

There are at least two main considerations when performing collocations. The first is the sampling pattern of the instrument. Measurements may be dense and regular, such as those from a cross-track scanning satellite instrument. In this case, some form of interpolation may be appropriate. On the other hand, the measurements may be sparse, such as a one-off field campaign designed for some particular validation exercise. The second consideration is the instrument footprint, which can range from something approaching a point measurement such as that of a radiosonde, or it might be several kilometers in diameter such as that of a satellite-mounted, microwave radiometer. In the latter case, it is appropriate to take into account the instrument antenna pattern when making comparisons with another instrument having both a smaller footprint and a denser sampling, that is, several measurements from the one instrument will fit into the footprint of the other.

Just as the instrument has a spatial footprint, it will also have a temporal footprint, often called the integration time. While the integration time is usually less than a second, which for meteorological applications is essentially instantaneous, there are many instances where some form of time averaging can considerably ease the collocation process.

The collocations will need to be screened based on both the time and length scales of the phenomenon of interest. This will further facilitate the collocation process since remote sensing and other measurement data is almost always binned in some way. Certain atmospheric phenomena such as clouds or convection are quite transient so that we need not consider collocations with a time error of more than an hour or so. Sea ice, on the other hand, moves and evolves quite slowly, so that measurements separated by as much as a day or more might still be useful.

Satellites

The satellites that most concern us are those with a low-Earth, polar orbit since geostationary satellites view the same point throughout their lifetime. The diagram shows measurements from AMSU-B instruments mounted on three satellites over a period of 12 hours. This illustrates both the orbit path and the scan pattern which runs crosswise. Since the orbit of a satellite is deterministic, barring orbit maneuvers, we can predict the location of the satellite at a given time and, by extension, the location of the measurement pixels. In theory, collocations can be performed by inverting the determining equations starting from the desired time period. In practice, partially processed data (usually referred to as level 1b, 1c or level 2) contain the coordinates of each of the measurement pixels and it is common to simply feed these coordinates to a nearest neighbor search. As mentioned previously, the satellite data is always binned in some manner. At minimum, the data will be arranged in swaths extending from pole to pole. The swaths will be labelled by time period and the approximate location known.

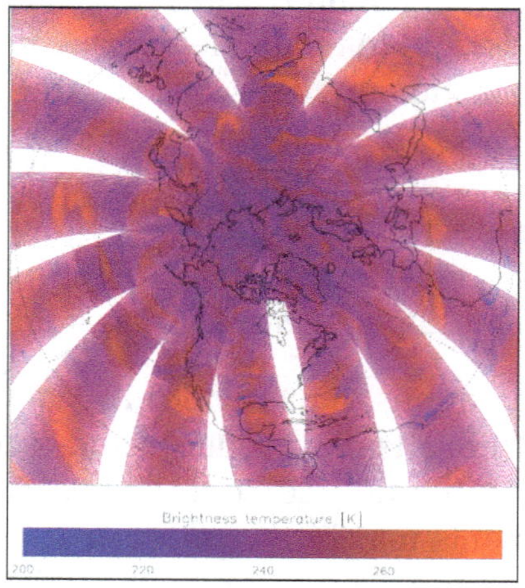

Polar-stereographic projection showing 12 hours of measurements from three AMSU-B instruments.

Radiosondes

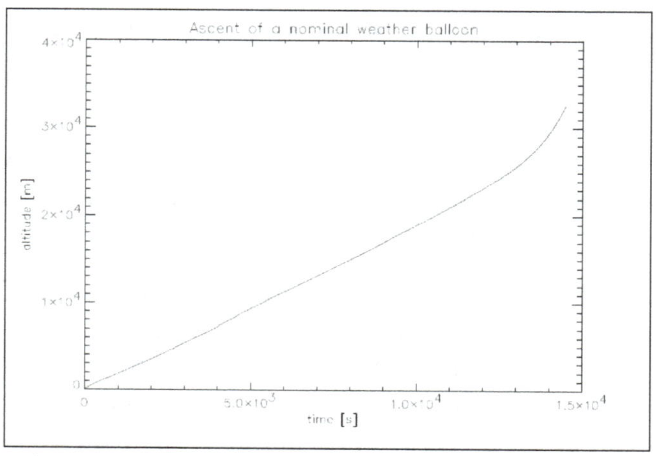

Ascent of a weather balloon launched from the Polarstern research vessel.

Radiosondes are particularly important for collocation studies because they measure atmospheric variables more accurately and more directly than satellite or other remote-sensing instruments. In addition, radiosonde samples are effectively instantaneous point measurements. One issue with radiosondes carried aloft by weather balloons is balloon drift. In, this is handled by averaging all the satellite pixels within a 50 km radius of the balloon launch.

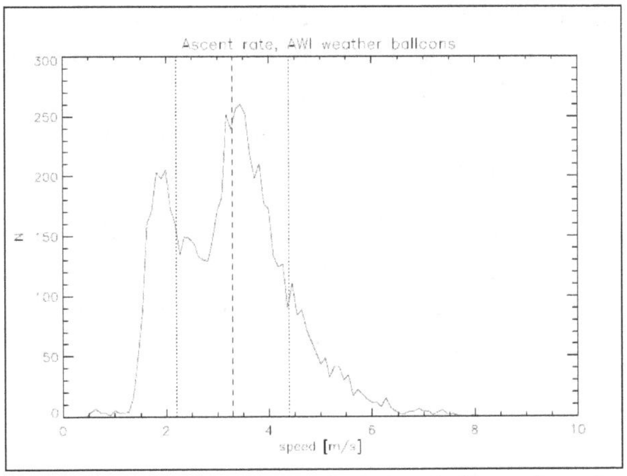

Histogram of ascent rates of weather balloons launch from the Polarstern research vessel.

If high-resolution sonde data, which normally has a constant sampling rate or includes the measurement time, is used, then the lateral motion can be traced from the wind data. Even with low-resolution data, the motion can still be approximated by assuming a constant ascent rate. Excepting a short bit towards the end, the linear ascent can be clearly seen in the figure above. We can show that the ascent rate of a balloon is given by the following equation:

$$v = \sqrt{\frac{gkh(1 - R_a / R_s)}{c_D}}$$

where g is gravitational acceleration, k relates the height, h, and surface area, A, of the balloon to its volume: $V = khA$; R_s is the equivalent "gas constant" of the balloon, R_a is the gas constant of the air and c_D is the drag coefficient of the balloon. Substituting some sensible values for each of the constants, k=1. (the balloon is a perfect cylinder), h=2. m, c_D = 1. and R_a is the gas constant of helium, returns an ascent rate of 4.1 m/s. Compare this with the values shown in the histogram which compiles all of the radiosonde launches from the Polarstern research vessel over a period of eleven years between 1992 and 2003.

Interpolation

For gridded data such as assimilation or reanalysis data, interpolation is likely the most appropriate method for performing any type of comparison. A specific point in both physical position and time is easy to locate within the grid and interpolation performed between the nearest neighbors. Linear interpolation (bilinear, trilinear etc.) is the most common, though cubic is used as well but is probably not worth the extra computational overhead. If the variable of interest has a relatively smooth rate of change (temperature is a good example of this because it has a diffusion mechanism, radiative transfer, not available to other atmospheric variables), then interpolation can eliminate much of the error associated with collocation.

Interpolation may also be appropriate for many types of satellite instruments, for instance a cross-track scanning instrument like Landsat. In data derived from the Advanced Microwave Sounding Unit (AMSU) are interpolated (although not for the purposes of collocation) using a slight variation of trilinear interpolation. Since measurements within a single scan track are laid out in an approximately rectangular grid, bilinear interpolation can be performed. By searching for the nearest overlapping scan track both forwards and backwards in time, the spatial interpolates can then be interpolated in time. This technique works better with derived quantities rather than raw brightness temperatures since the scan angle will already have been accounted for.

For instruments with a more irregular sampling pattern, such as the Advanced Microwave Scanning Radiometer-EOS (AMSR-E) instrument which has a circular scanning pattern, we need a more general form of interpolation such as kernel estimation. A method commonly used for this particular instrument, as well as SSM/I, is a simple daily average within regularly gridded, spatial bins.

Trajectories

To collocate measurements of a medium-to long-lived atmospheric tracer with a second instrument, running trajectories can considerably improve the accuracy. It also simplifies the analysis somewhat: a trajectory is run both forwards and backwards from the measurement location and between the desired time window. Note that the acceptable time window has now become longer because the error from transport induced changes in the tracer is removed: the tracer lifetime would be a good window to use. Since the trajectories provide a location for every point in time within the time window, there is no need to check multiple measurements from the second instrument. Every time within the trajectory is checked for the distance criterion but within a very narrow window. Alternatively, the exact times of the measurements for the second instrument are interpolated within the trajectory. Only the smallest distance error below the threshold is used and the distance criterion can be made smaller as a consequence.

Pol-Ice Campaign

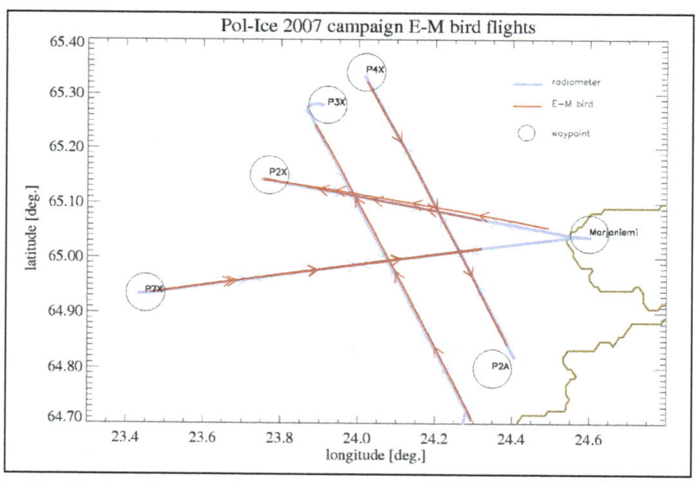

Map of E-M Bird flights from Pol-Ice campaign along with coincident EMIRAD flights.

Collocations of sea ice thickness and brightness temperatures taken during the Pol-Ice Campaign are an excellent example since they illustrate many of the most important principles as well as demonstrating the necessity of taking into account the individual case. The Pol-Ice campaign was conducted in the N. Baltic in March 2007 as part of the SMOS-Ice project in preparation for the launch of the Soil Moisture and Ocean Salinity satellite. Because of the low frequency of the SMOS instrument, it is hoped that it will render information on sea ice thickness, therefore the campaign comprised measurements of both sea ice thickness and emitted brightness temperature. Brightness temperatures were measured with the EMIRAD L-band microwave radiometer carried on board an airplane. Ice thickness was measured with the E-M Bird ice thickness meter which was carried by a helicopter. The E-M Bird measures ice thickness with a combination of inductance measurements to determine the location of the ice-water interface and a laser altimeter to measure the height of the ice surface. The map above shows the flight tracks of both instruments which were approximately coincident but obviously subject to pilot error.

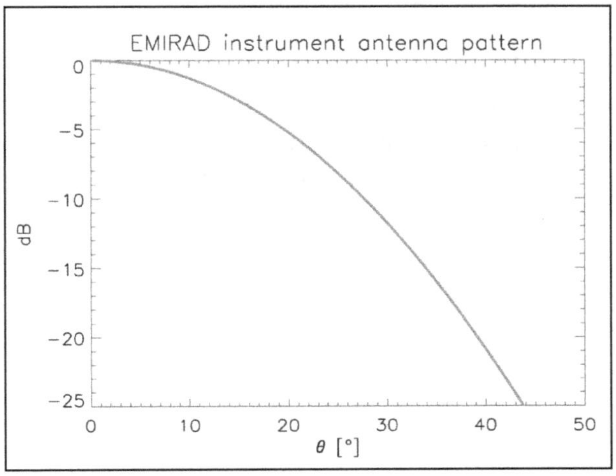

EMIRAD antenna response pattern.

Since the flight paths of both aircraft were approximately linear, the first step in the collocation process was to convert all the coincident flights to Cartesian coordinates with the x-axis being lateral distance and the y-axis transverse distance. In this way, collocations can be performed in two ways: crudely, by matching only the x distances, and more precisely by matching both coordinates.

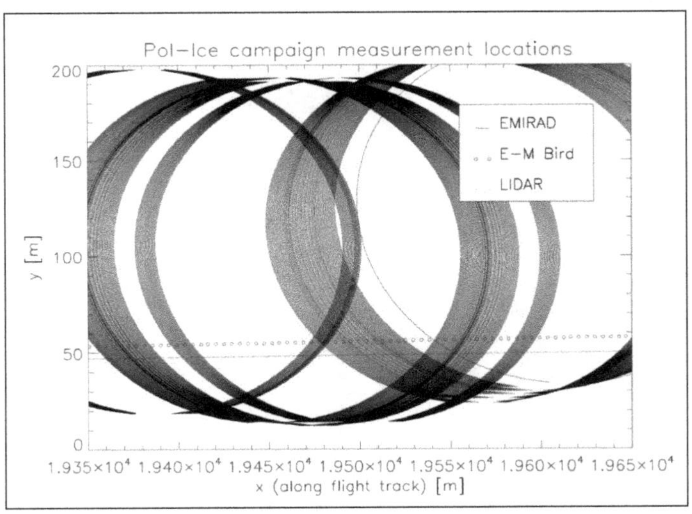

Relative measurement locations from P4X to P2A flight track: see above map. EMIRAD footprints represent the Gaussian standard-deviation, not FWHM.

More importantly, the footprint size of the radiometer is many times larger than that of the E-M Bird meter. The figure to the left shows the antenna response function for the radiometer. The full width at half maximum is 31 degrees. Since the aircraft was flying at approximately 500 m, this translates to a footprint size of 200 m or more. Meanwhile, the footprint size of the E-M Bird was roughly 40 m with a sample spacing of only 2 to 4 m. Rather than looking to nearest neighbors, which would have produced poor results, a weighted average of the thickness measurements was performed for each radiometer measurement. Weights were calculated based on the radiometer response function which is almost a perfect Gaussian up to about 45 degrees. Points could be excluded based on distance along the flight path. For validation of sea ice emissivity forward model calculations, this was further refined by performing an emissivity calculation for each thickness measurement and averaging over the radiometer footprint.

The figure below illustrates relative measurement locations from each of the instruments used in the Pol-Ice campaign. Two overpasses are shown: one from the airplane carrying the EMIRAD radiometer and one from the helicopter carrying the E-M Bird instrument. The x-axis is along the line of the flight path. EMIRAD footprints are drawn with lines, E-M Bird inductance measurements are represented by circles and LIDAR measurements with dots.

Types and Applications of Actuators 5

- Linear Solenoid Actuator
- Pneumatic and Hydraulic Actuators
- Carbon Nanotube Actuators
- Piezoelectric Actuator
- Helical Band Actuator
- MEMS Magnetic Actuator
- Mini/Micro Shape Memory Actuators
- Industrial Applications of Electric Linear Actuator

There are numerous types of actuators that are applied in different areas. Some of them include linear solenoid actuator, pneumatic and hydraulic actuators, carbon nanotube actuators, piezoelectric actuator, helical band actuator, MEMS magnetic actuator and mini/micro shape memory actuators. The topics elaborated in this chapter will help in gaining a better perspective about these types and applications of actuators.

Linear Solenoid Actuator

The linear solenoid works on the same basic principal as the electromechanical relay and just like relays, they can also be switched and controlled using transistors or MOSFET's. A "Linear Solenoid" is an electromagnetic device that converts electrical energy into a mechanical pushing or pulling force or motion.

Linear solenoid's basically consist of an electrical coil wound around a cylindrical tube with a ferro-magnetic actuator or "plunger" that is free to move or slide "IN" and "OUT" of the coils body. Solenoids can be used to electrically open doors and latches, open or close valves, move and operate robotic limbs and mechanisms, and even actuate electrical switches just by energising its coil.

Linear Solenoid

Solenoids are available in a variety of formats with the more common types being the linear solenoid also known as the linear electromechanical actuator, (LEMA) and the rotary solenoid.

Both types of solenoid, linear and rotational are available as either a holding (continuously energised) or as a latching type (ON-OFF pulse) with the latching types being used in either energised or power-off applications. Linear solenoids can also be designed for proportional motion control were the plunger position is proportional to the power input.

When electrical current flows through a conductor it generates a magnetic field, and the direction of this magnetic field with regards to its North and South Poles is determined by the direction of the current flow within the wire. This coil of wire becomes an "Electromagnet" with its own north and south poles exactly the same as that for a permanent type magnet.

The strength of this magnetic field can be increased or decreased by either controlling the amount of current flowing through the coil or by changing the number of turns or loops that the coil has. An example of an "Electromagnet" is given below.

Magnetic Field Produced by a Coil

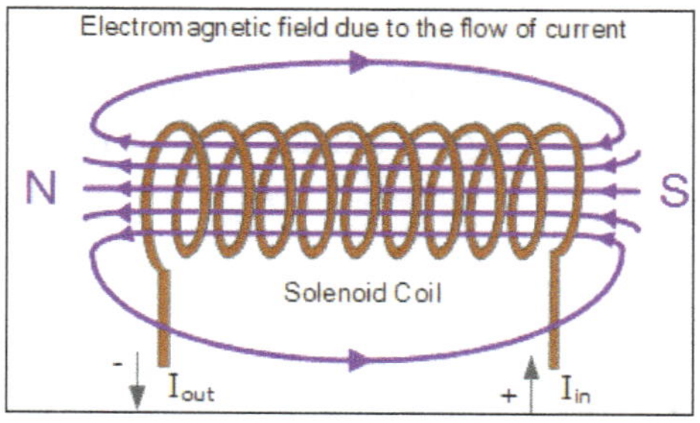

When an electrical current is passed through the coils windings, it behaves like an electromagnet and the plunger, which is located inside the coil, is attracted towards the centre of the coil by the magnetic flux setup within the coils body, which in turn compresses a small spring attached to one end of the plunger. The force and speed of the plungers movement is determined by the strength of the magnetic flux generated within the coil.

When the supply current is turned "OFF" (de-energised) the electromagnetic field generated previously by the coil collapses and the energy stored in the compressed spring forces the plunger back out to its original rest position. This back and forth movement of the plunger is known as the solenoids "Stroke", in other words the maximum distance the plunger can travel in either an "IN" or an "OUT" direction, for example, 0 – 30mm.

Linear Solenoid Construction

This type of solenoid is generally called a Linear Solenoid due to the linear directional movement and action of the plunger. Linear solenoids are available in two basic configurations called a "Pull-type" as it pulls the connected load towards itself when energised, and the "Push-type" that act in the opposite direction pushing it away from itself when energised. Both push and pull types are generally constructed the same with the difference being in the location of the return spring and design of the plunger.

Pull-type Linear Solenoid Construction

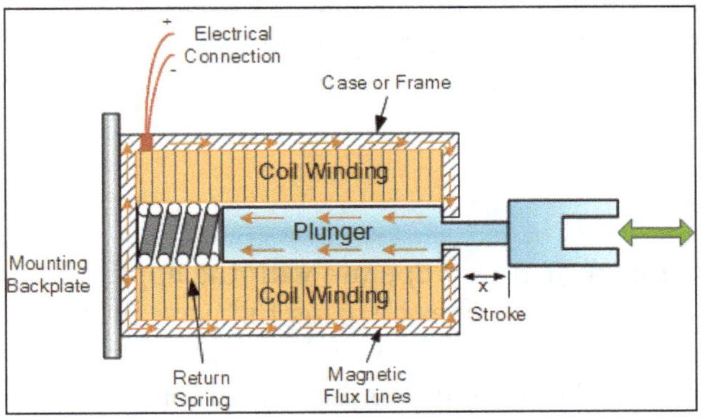

Linear solenoids are useful in many applications that require an open or closed (in or out) type motion such as electronically activated door locks, pneumatic or hydraulic control valves, robotics, automotive engine management, irrigation valves to water the garden and even the "Ding-Dong" door bell has one. They are available as open frame, closed frame or sealed tubular types.

Rotary Solenoids

Most electromagnetic solenoids are linear devices producing a linear back and forth force or motion. However, rotational solenoids are also available which produce an angular or rotary motion from a neutral position in either clockwise, anti-clockwise or in both directions (bi-directional).

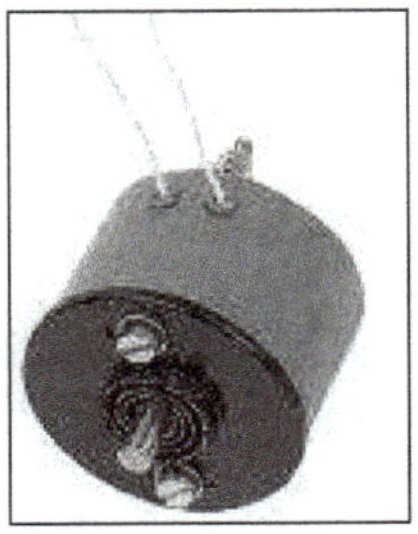

Rotary Solenoid

Rotary solenoids can be used to replace small DC motors or stepper motors were the angular movement is very small with the angle of rotation being the angle moved from the start to the end position.

Commonly available rotary solenoids have movements of 25, 35, 45, 60 and 900 as well as multiple movements to and from a certain angle such as a 2-position self restoring or return to zero rotation, for example 0-to-90-to-00, 3-position self restoring, for example 00 to +450 or 00 to -450 as well as 2-position latching.

Rotary solenoids produce a rotational movement when either energised, de-energised, or a change in the polarity of an electromagnetic field alters the position of a permanent magnet rotor. Their construction consists of an electrical coil wound around a steel frame with a magnetic disk connected to an output shaft positioned above the coil.

When the coil is energised the electromagnetic field generates multiple north and south poles which repel the adjacent permanent magnetic poles of the disk causing it to rotate at an angle determined by the mechanical construction of the rotary solenoid.

Rotary solenoids are used in vending or gaming machines, valve control, camera shutter with special high speed, low power or variable positioning solenoids with high force or torque are available such as those used in dot matrix printers, typewriters, automatic machines or automotive applications etc.

Solenoid Switching

Generally solenoids either linear or rotary operate with the application of a DC voltage, but they can also be used with AC sinusoidal voltages by using full wave bridge rectifiers to rectify the supply which then can be used to switch the DC solenoid. Small DC type solenoids can be easily controlled using Transistor or MOSFET switches and are ideal for use in robotic applications.

However, as we saw previously with electromechanical relays, linear solenoids are "inductive" devices so some form of electrical protection is required across the solenoid coil to prevent high back emf voltages from damaging the semiconductor switching device. In this case the standard "Flywheel Diode" is used, but you could equally use a zener diode or small value varistor.

Switching Solenoids using a Transistor

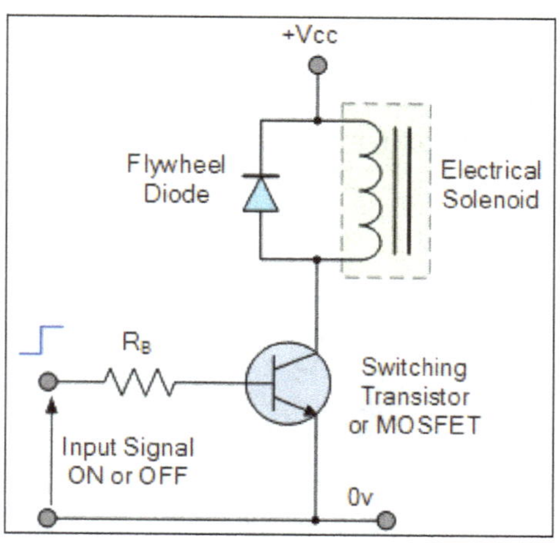

Reducing Energy Consumption

One of the main disadvantages of solenoids and especially the linear solenoid is that they are "inductive devices" made from coils of wire. This means that the solenoid coil converts some of the electrical energy used to operate them into "HEAT" due to the resistance of the wire.

In other words when connected for long periods of time to an electrical supply they get hot!, and the longer the time that the power is applied to a solenoid coil, the hotter the coil will become. Also as the coil heats up, its electrical resistance also changes allowing more current to flow increasing its temperature.

With a continuous voltage input applied to the coil, the solenoids coil does not have the opportunity to cool down because the input power is always on. In order to reduce this self generated heating effect it is necessary to reduce either the amount of time the coil is energised or reduce the amount of current flowing through it.

One method of consuming less current is to apply a suitable high enough voltage to the solenoid coil so as to provide the necessary electromagnetic field to operate and seat the plunger but then once activated to reduce the coils supply voltage to a level sufficient to maintain the plunger in its seated or latched position. One way of achieving this is to connect a suitable "holding" resistor in series with the solenoids coil.

Reducing Solenoid Energy Consumption

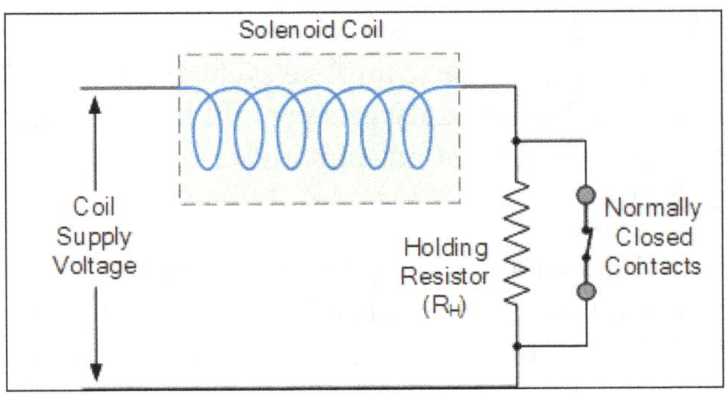

Here, the switch contacts are closed shorting out the resistance and passing the full supply current directly to the solenoid coils windings. Once energised the contacts which can be mechanically connected to the solenoids plunger action open connecting the holding resistor, RH in series with the solenoids coil. This effectively connects the resistor in series with the coil.

By using this method, the solenoid can be connected to its voltage supply indefinitely (continuous duty cycle) as the power consumed by the coil and the heat generated is greatly reduced, which can be up to 85 to 90% using a suitable power resistor. However, the power consumed by the resistor will also generate a certain amount of heat, I^2R (Ohm's Law) and this also needs to be taken into account.

Solenoid Duty Cycle

Another more practical way of reducing the heat generated by the solenoids coil is to use an "intermittent duty cycle". An intermittent duty cycle means that the coil is repeatedly switched "ON" and "OFF" at a suitable frequency so as to activate the plunger mechanism but not allow it to de-energise during the OFF period of the waveform. Intermittent duty cycle switching is a very effective way to reduce the total power consumed by the coil.

The Duty Cycle (%ED) of a solenoid is the portion of the "ON" time that a solenoid is energised and is the ratio of the "ON" time to the total "ON" and "OFF" time for one complete cycle of operation. In other words, the cycle time equals the switched-ON time plus the switched-OFF time. Duty cycle is expressed as a percentage, for example:

$$\text{Duty Cycle} = \frac{\text{"ON" time}}{\text{"ON" time} + \text{"OFF" time}} \times 100\%$$

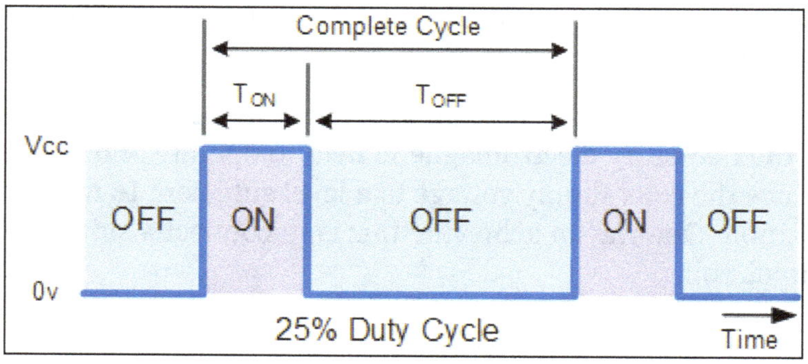

Then if a solenoid is switched "ON" or energised for 30 seconds and then switched "OFF" for 90 seconds before being re-energised again, one complete cycle, the total "ON/OFF" cycle time would be 120 seconds, (30+90) so the solenoids duty cycle would be calculated as 30/120 secs or 25%. This means that you can determine the solenoids maximum switch-ON time if you know the values of duty cycle and switch-OFF time.

For example, the switch-OFF time equals 15 secs, duty cycle equals 40%, therefore switch-ON time equals 10 secs. A solenoid with a rated Duty Cycle of 100% means that it has a continuous voltage rating and can therefore be left "ON" or continuously energised without overheating or damage.

Pneumatic and Hydraulic Actuators

Pneumatic Actuators

Physical processes proceeding in pneumatic drives are submitted to the gas laws. The gas laws are a set of laws that describe the relationship between thermodynamic temperature (T), pressure (P) and volume (V) of gases. Three of these laws, Boyle's law, Charles's law, and Gay-Lussac's law, may be combined to form the combined gas law:

$$\frac{P_1 V_1}{T_1} = \frac{P_2 V_2}{T_2}$$

which with the addition of Avogadro's law later gave way to the ideal gas law. Other important gas laws include Dalton's law of partial pressures. The kinetic theory of gases, Graham's law of effusion and root mean square velocity explains how individual molecules act in a gas and their relation to pressure, volume, and temperature.

A gas that obeys these gas laws is known exactly as an ideal gas (or perfect gas). An ideal gas does not exist; however, some gases follow the laws more closely than the others in given standard conditions.

The most important gas law is the ideal gas law, which states that: $PV = nRT$. Other gas laws, such as van der Waals equation, seek to correct the ideal gas laws to reflect the behaviour of actual gases. Van der Waals equation alters the ideal gas law to reflect how actual gases function using a series of calculated values called van der Waals constants. It is also possible to apply Boltzmann's analysis to determine further information about gases. With the equation:

$$\frac{1}{2}m < c^2 > = \frac{3}{2}kT$$

detailed information about the speeds of the atoms in a gas, the kinetic energy, and the temperature conditions under which it exists can be gleaned.

Classification of Pneumatic Actuators

A set of devices into with one or more pneumoengines, which are determined to start mech- anisms or some other objects by means of pressed working gas is called pneumatic actuator, or pneumoactuator.

The devices intended for transformation of potential and kinetic energy of the stream of compressed gas in mechanical energy of the output link that can be, for example, a rod of the piston, a shaft of the turbine or the case of the jet device is called pneumatic engines of the automated actuator.

All pneumatic actuators can be subdivided into the following types:

- Diaphragm pneumatic actuators.
- Pneumatic power cylinders.
- Gas-engine pneumatic actuators.
- Turbine pneumatic actuators.
- Jet-stream pneumatic actuators.
- Pneumomuscles.
- Combined pneumatic actuators.

The principle of transformation of potential or kinetic energy of the gas stream into mechanical energy of the engine output link of the engine provide the base for division into types.

Diaphragm Pneumatic Actuators

Diaphragm pneumatic actuators include membrane and sylphon pneumoactuators. Potential energy of gas stream, i.e. energy of static pressure is used in diaphragm pneumatic actuators. The difference of static pressure in receivers of the actuator transformes into the effort on the output element of the drive - the rod. Pneumatic actuators can be executed both on one-sided (not reversive), and on double-sided (reversive) scheme.

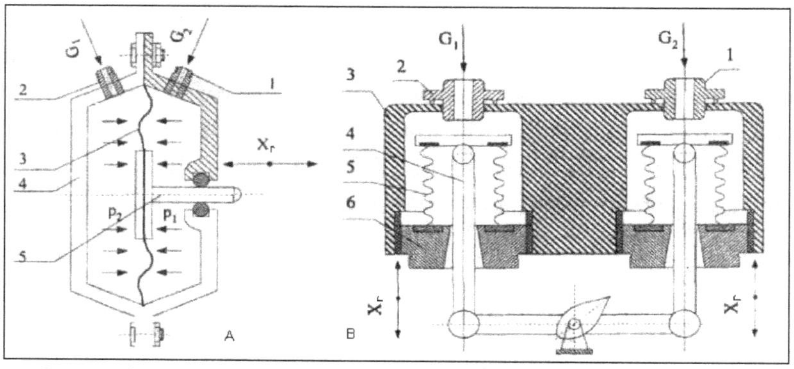

A - The membrane pneumatic actuator; B - The sylphon pneumatic actuator.

The basic scheme of the membrane pneumatic actuator is shown in figure. Work of the pneumatic actuator consists in moving of a rod under influence of the difference of pressure p_1 p_2 in cavities that is formed due to the difference of gas charges G_1 G_2. The sylphon pneumatic engine functions by the same principle.

Pneumatic Power Cylinders

Air, or pneumatic cylinders are devices that convert power of compressed air into mechanical energy. This mechanical energy produces linear or rotary motion. In this way, the air cylinder functions as the actuator in the pneumatic system, so it is also known as a pneumatic linear actuator.

Devices with forward linear movement are divided into single-acting and double-acting pneumocylinders, with rod and rodless. Rod pneumocylinders in turn can have a through-pass or a no-go rod. Devices with rotary (rotational) movement are divided into pneumocylinders with rotary movement of the output link and rotational pneumocylinders.

Structurally pneumatic power cylinders can be piston, membrane, sylphon and hose.The air cylinder consists of steel or stainless steel piston, piston rod, cylinder barrel and end covers.

In piston pneumatic cylinders, as well as in diaphragm, potential energy of the compressed gas is used, but presence of the piston with mobile condensation allows to reach big movings of the output link.

As compressed air moves into the cylinder, it pushes the piston along the length of the cylinder. Compressed air or the spring, located at the rod end of the cylinder, pushes the piston back.

For compression a cuff from elastic material is installed into the rod in a cover of the cylinder. The bush serving as direction for the rod, allows to increase the allowable cross- section effort of the loading on to the rod, especially for big lengths. The ring that prevents dust and firm particles hitting the cylinder is inserted into the cover, in front of the bush. Volume changing of the working cavities is into proportionally to the moving of the rod. Thus, the additional volume can be reductioned to a minimum, determined only by the volume of the pipelines.

Some form of cushioning is normally required to reduce the rate of the cylinder travel before the piston strikes the end cover. Reducing the piston velocity at the end of its travel lowers the stresses on the cylinder while reducing vibration in the structure of which it is part. Efficient cushioning is usually necessary in applications, in which precision is required. Basically, this problem can be solved in one of three ways: by means of simple impact cushioning, by pneumatic cushioning or by fitting shock absorbers.

Valves control the flow of compressed air to the cylinder.Different types of cylinders include standard, short-run, flat, rodless, rotary cylinders. The two basic types of standard air cylinders, based on their performance, are as follows: single acting and double acting. The single acting cylinder is able to perform an operating motion in only one direction. Ram cylinder - a cylinder in which the movable element is the piston rod. A double acting pneumatic cylinder — has two-directed powered motion in, with pressure on both sides. When a cylinder is pushed out in one direction, compressed air moves it back in the opposite direction. Air lines running into both ends of the cylinder supply the compressed air. The flow of the compressed air is controlled with valves both in single- and double-acting cylinders.

Spring return cylinder - a cylinder in which a spring returns the piston assembly. A single acting - spring return air cylinder - has air pressure on one side of the piston flange, supplying force and motion, and the spring supplying the return force after the pressure release. Single acting cylinders require approximately a half of the amount of the air used by a double acting cylinder for a single operating cycle.

Double-acting double-rod cylinder, double-acting cylinder with a piston rod extending form each end. The piston rods are connected with the same piston. Double rod cylinders provide equal force and speed in both directions. Double acting with magnetic piston according to DIN ISO 6431/VDMA 24562 (cylinders with piston rod on both sides).

Pneumatic Power Cylinders.

An extremely flat design makes it possible to position the ball-bearing system between both pistons. This series is a concept with clear advantages. Thanks to the double-acting principle, both piston rods and the guide system work together precisely and parallel. The ball-bearing circulating guide guarantees a longer service life when operated vertically and the double pistons provide a high level of force, in spite of the flat design. This flexible and light design, as well as precise movements, make it possible to use the slide in dynamic applications. Numerous holes also provide a wide variety of mounting options and facilitate machine layout.

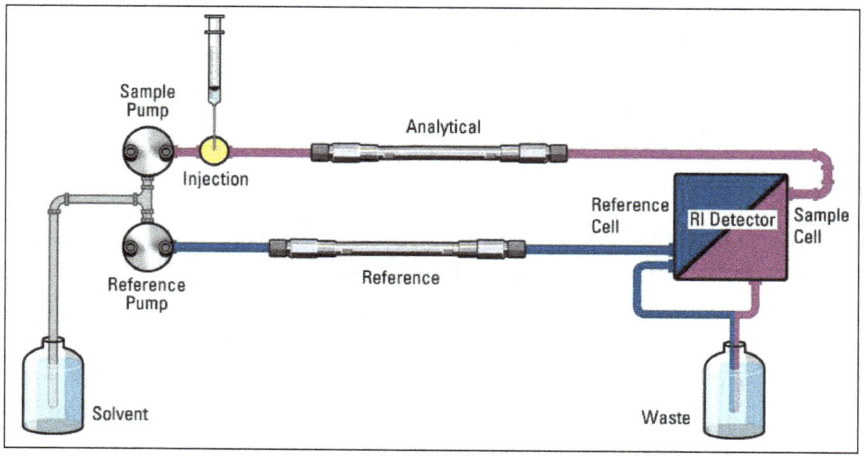

Another series is ideal for all applications that demand absolute precision and side load capacity. Compared to standard cylinders, cylinders from this series offer extremely precise movement, high side load capacities, and also torsion protection. This results in fewer outer guides to design and set up machines. It optimally designed for handling technology. Thanks to a wide range of technical features, the guided precision cylinder is the perfect product for automation tasks. For example: The next smallest cylinder fits directly on the front plate of the larger cylinder. The Top-Loader version even has an additional mounting surface on the top that can be used to attach extra components and objects. It runs smoothly and precisely thanks to efficient plain bearing guides or ball bushing guides, making separate guides unnecessary. Guide cylinders from this series are the top choice for the following applications:

- As a stop in a guide system.
- To fasten workpieces for machine processing or spot welding.
- As a support for a second axis.
- To hold mechanical grippers or suction cups for handling tasks.
- For pick and place systems.

The tandem-cylinder represents two cylinders of bilateral action with the general rod, incorporated in one case. By connecting 2, 3 or 4 cylinders with the same piston diameter and stroke in series, the force in the advance stroke (thrust) can be doubled, tripled or quadrupled in comparison to a single cylinder. Tandem-cylinders are applied where it is necessary to develop significant efforts at the limited cross-section sizes of the cylinder.

In the course of rodless cylinders three principles of constructive design were used: tape or rope cylinders, cylinders with condensation longitudinal of spline and cylinders with magnetic socket.

The output part of the rodless pneumocylinder is the carriage to which the loading is connected. The effort from the piston is transferred to the carriage through consolidation longitudinal spline. In the cylinder, along all length of the case there is a through slit. The carriage is rigidly connected with the piston. This connection is carried out with the help of the element sliding in spline of the case pipe of the cylinder. Consolidation spline is provided with a steel tape which adjoins to the internal side of the spline. Each cavity of the cylinder is hermetically sealed with the consolidation established on the piston. Between these consolidation the tape is bent and passes under the element connecting the piston with the carriage. The second tape consolidates spline outside, that protects the cylinder from the hit of pollution from the outside.

The rodless cylinder with consolidation longitudinal spline.

In comparison with the usual cylinder of double-action the rodless cylinder possesses smaller length of the design. Restrictions of cross-section loading on the working body, as against the roding cylinder, are removed. The rodless cylinder can be applied at courses up to 10 m.

The tape cylinder transfers force from the piston to the carriage with the help of the tape covering it. At an output from working volume of the cylinder the tape passes through consolidation in covers of the cylinder.

The cylinder with magnetic coupling of double action consists of cylindrical case, rodless piston and mobile external carriage sliding on the external surface of the case of the cylinder, having in cross-section section the form of a circle. On the piston and on the carriage the constant magnets, cooperating with each other themselves are placed, i.e. transfer of effort for moving loading from the piston to the carriage is carried out with the help of magnetic coupling. After the air have been submitted to the cylinder, the carriage and the piston move simultaneously. The cavity of the cylinder is tight and has no mobile consolidation adjoining to the environment. It completely excludes an opportunity of outflow of air from the cylinder. It is possible to relate pneumoengines of back and forth motion conditionally to the hose pneumocylinders, which are used in transporting mechanisms with significant movings, but with small moved weights. Rollers of the carriage swage the elastic hose. When the air is submitted to the end of the hose and the other end is connected-with the atmosphere while the control valve takes position A, the carriage starts to move together with the moved object due to the hose deformation caused by compressed air. Back-movement of the carriage takes place at switching the control valve in position B.

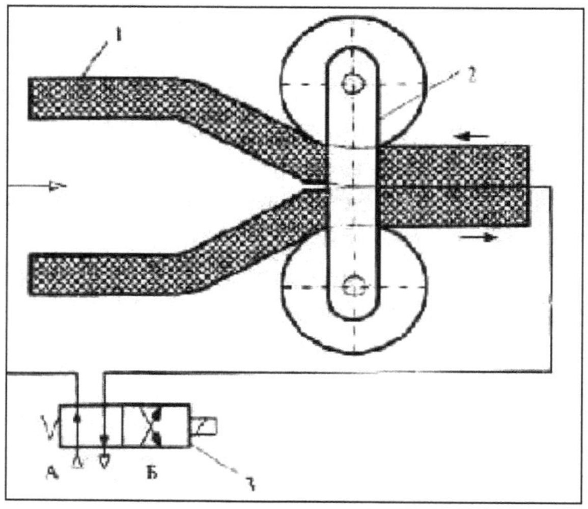

The hose pneumocylinder.

Rotary pneumoactuators basically are constructed on the principle of mechanical trans- formation of progress of the piston in rotary movement of a target link.

Pneumomuscles

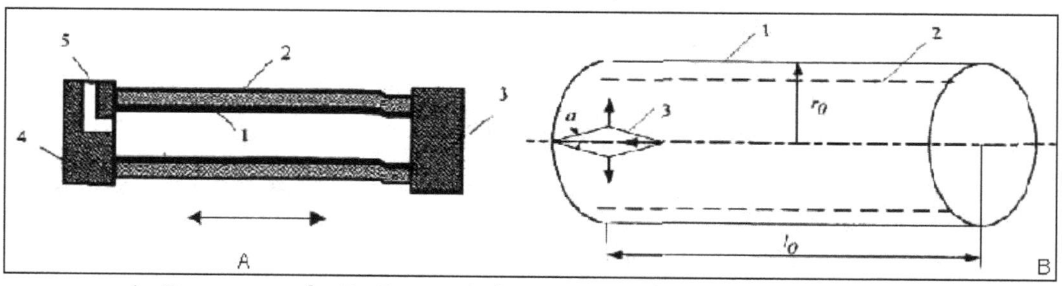

A - Pneumomuscle; B - Geometrical parameters of reduced pneumomuscles.

Pneumomuscles are linear pneumatic drives which can reproduce the movements similar to natural muscles. In comparison with power cylinders they develop the big initial efforts to the same working diameters. We distinguish lengthen and reduced pneumomuscles. Character of movement is defined with geometry of a braid of a pneumomuscle. The scheme of a lengthen pneumomuscle is given in figure.

The internal elastic tube has the braid with three-dimensional cellular structure. Cells have the form of rhomboid. When the channel pressure moves into the feed, the tube starts to extend in its both sides, as it is shown in figure, if the both covers are not fixed. If one of the covers is fixed, an output part of this drive will be the other cover. The design has no moving mechanical parts and it is reliable. The working course of the output link is up to 25 percents from the length of the pneumomuscle.

The effort developed by the drive has the maximal size at the beginning of the working course and further it practically decreases linearly. It allows to realize significant acceleration of the output link in a combination withthe smooth and exact approach to the desirable final position.

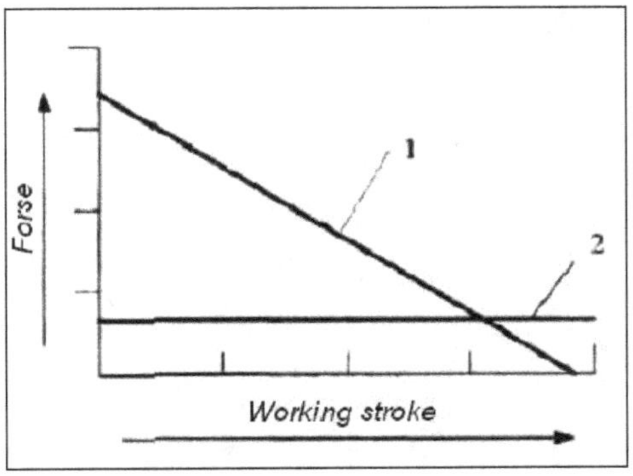

Comparative characteristics.

Unlike lengthening pneumomuscles reduced pneumomuscles have a braid in the form of cells, which form is a grid from pantographs and transform the pressure inside the elastic tube when it reduces into length.

Due to symmetry of the braid the pneumomuscle always keeps the cylindrical form. The initial angle a is defined as the angle between the longitudinal axis of the pneumomuscle and the side of the cell when the working pressure is absent. Parameters ro and lo are initial radius and length of a pneumomuscle.

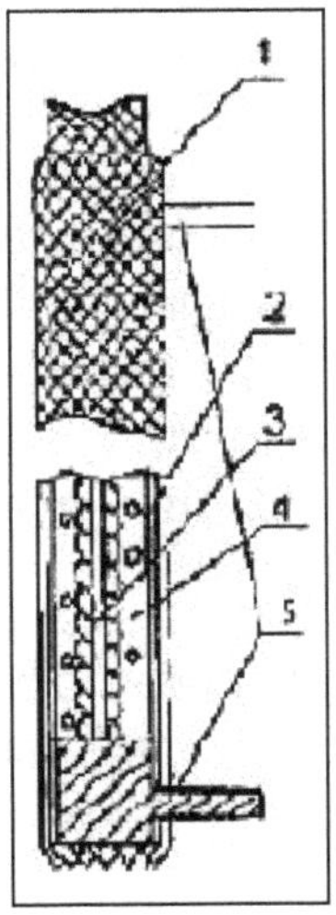

The artificial muscle.

The variant of the pneumatic artificial muscle, in which the process of gas pressing takes place directly in the tube as a result of heating by an electric current, is shown in figure. When the electric current passes through the thermoelement 3 special filler 4 is heated up and eduse gas. As a result the pressure grows inside the tube and it blows out and reduces its length. After switching-off of the current filler cools and again absorbs the educed gas, the muscle comes back to the initial condition.

Length of the muscle is 150-300`1`1, diameter of 3-5 mm. Size of reduction - 10-15 percents. Developed effort - 1-3 ´I, at weight of a muscle - units of grammes. The main drawback of this muscle - very low speed: time of compression is more then 30 seconds, and it is 2-3 times as mush as the return relaxation. Other variant of the similar pneumatic muscle with thermal influence - filling of the tube by low-boiling liquid while the external sourse of heat influences on it.

The Combined Pneumatic Actuators

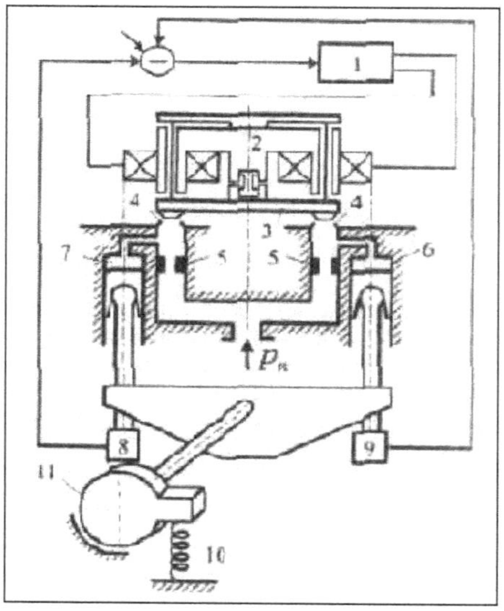

The electropneumatic actuator.

In particular, electropneumatic actuators considered to be combined drives. In electrop- neumatic drives executive engines, electromechanical converters and amplifiers are applied. For example, we shall consider the circuit of an electropneumatic follower drive, in which hot gas serves as a work environment.

The drive consume gas from the source of constant pressure. Basic elements of the drive are the amplifier 1 of direct current, the electromechanical converter 2 with Ø-shaped stator, choker 3 that is elastic fixed on stator, nozzles 4, throttles 5, executive pneumocylinders 6 and 7, sensors 8 and 9 of feedback by speed and on position of a conducted link - the shaft connected to loading. Item loading is given by a spring 10, and inertial loading - a flywheel 11. Besides it is supposed, that the loading from hydraulic friction is also created.

In the given drive, the executive engines are pneumocylinders that are connected directly to the channels of the pneumoamplifier with managing element of nozzle- choker type, therefore due to the use of energy produced by gas which is under pressure only one step of amplification is carried out.

Choker 3, controlled by the electromechanical converter, carries out regulation of gas streams through out the constant throttles 5.

Recuperation of Energy in Drives

Drives of the robots work basically in transitive cyclic modes dispersal - braking type. Therefore as one of the most important ways to economise on energy at them is to use the idea of energy recuperation. It is especially important for mobile robots with independent power supply. Moreover the speed of operation also raises frequently. Recuperation of energy is based on its storage during braking and feedback at the subsequent acceleration. There are two basic ways of such storage of energy: storage of mechanical energy (with the help of flywheels, springs) and electric (in accumulators, condensers, inductive coils).

A principle of mechanical energy recuperation with the help of springs has got application in Russian cyclic industrial robots. That is why they have very good power characteristics. The spring that provides cyclic movement of the manipulator in the mode of resonant not fading fluctuations with zero speed in extreme points is installed into the cyclic drive. The drive engine carries out the energy charging in the middle of the way at the maximal speed of movement, replenishing the energy loss by the drive while performing its work. Thus, in the final points of movement there is no impact on useless dispersion of the kinetic energy saved up by a drive. Thus, it is possible to reduce energy consumption by 3-4 times and the capacity of the engine can be accordingly reduced. The similar effect can be reach in drives of manipulators' claw devices.

Advantages

- Simplicity of realization relatively to small back and forth motions;
- Sophisticated transfer mechanisms are not required;
- Low cost;
- High speed of moving;
- Ease at reversion movements;
- Tolerance to overloads, up to a full stop;
- High reliability of work;
- Explosion and fire safety;
- Ecological purity;
- Ability to accumulation and transportation.

Disadvantages

- Compressibility of the air;
- Impossibility to receive uniform and constant speed of the working bodies movement;
- Difficulties in performance at slow speed;
- Limited conditions - use of compressed air is beneficial up to the definite values of pressure;
- Compressed air requires good preparation.

Hydraulic Actuators

The Laws of Work Hydraulic Actuators

Liquid is the physical body possessing fluidity. Fluidity is understood as easy mobility of particles, i.e.(that is) ability beyond all bounds to be deformed and come in movement under action of insignificant forces.

Due to fluidity liquids practically without efforts get the form given by space, and in balanced condition do not perceive tangents strain.

Liquids show resistance to shift in direct dependence on speed of shift deformation and to compression. They don't practically resist to stretching except in cases under special conditions, for example at all-round negative pressure or at attempt to tear off the piston from liquid in the cylinder.

However liquids considered technically pure, in practical conditions contain small impurity and dissolved air, therefore they are not capable to maintain even insignificant stretching strain.

Taking into account mechanical properties we can distinguish two types of liquid: low- compressible (dropping) liquids which insignificantly change the volume when temperature and pressure change, and compressed (gaseous). The difference of mechanical properties of these two types of liquid is insignificant, if it is possible to neglect compressibility of gases namely when differences of temperatures and pressure are small, and speed of movement less than 70 km/s. In these cases the laws received for drop liquids, can be frequently fair for gases. From physics approach the difference between these kinds of liquid is significant. Drop liquids have quite certain volume and consequently fill in only that part of space which is equal to their volume. They can form a boundary free surface, and at insignificant volumes are capable to keep the drop shape.

The phenomenon arising in the moving liquid when pressure decline up to elasticity of saturated vapor and an ambient temperature is named cavitation. Cavitation is accompanied by formation of steam-to-gas bulbs that move with a stream of liquid in area with higher pressure, slam and radiate shock wave. Slamming bulbs near borders of the current can course the destruction of firm surfaces.

Cavitation can arise at low pressure in pipelines, pumps - everywhere where the stream of liquid is exposed to bends, to narrowings expansion (valves, throttles, etc.) followed by.

As a rule, cavitation is the undesirable phenomenon, and should be eliminated from hy- drosystems because cavitation can course erosion of surfaces, destruction of elements of hydro- machines and hydrosystems. It can increase the resistance of pipelines and sharp by reduce efficiency of hydromachines.

Boiling and cavitation breaks homogeneity of liquids and limits a scope of the laws based on homogeneous nature of the liquid medium.

Silting of narrow cracks and backlashes owing to adsorption (adjournment) of polar-acting molecules of liquid on their walls is called obliteration. The layer of liquid formed at walls gets properties of quasisolid state which viscosity differs from viscosity of liquid. As a result, the part of a boundary layer sticks to a surface of a crack. The thickness of this layer for oils is 4...5 microns, which can essentially reduce a cross-section of slot-hole channels and backlashes or even overgrow them completely.

In fluid dynamics, Bernoulli's equation describes the behavior of fluid moving along the streamline. There are typically two different formulations of the equations; one applies to incompressible flow and the other applies to compressible flow.

The original form, for incompressible flow in a uniform gravitational field (such as on Earth), is:

$$\frac{v^2}{2} + gh + \frac{p}{\rho} = const$$

The equation is named after Daniel Bernoulli although it was first presented in the above form by Leonhard Euler.

The second, more general form of Bernoulli's equation may be written for compressible fluids, in which case, following the streamline:

$$\frac{v^2}{2} + \varphi + \omega = const$$

The constant on the right hand side is often called the Bernoulli constant and denoted b. For steady inviscid adiabatic flow with no additional sources or sinks of energy, b is constant along any given streamline. More generally, when b may vary along streamlines, it still proves a useful parameter, related to the "head" of the fluid.

When there are shock waves present, in a reference frame moving with a shock, many of the parameters in the Bernoulli equation suffer abrupt changes in passing through the shock. The Bernoulli parameter itself, however, remains unaffected. An exception to this rule is radiative shocks, which violate the assumptions leading to the Bernoulli equation, namely the lack of additional sinks or sources of energy.

Structure of Hydraulic Cylinders

Hydraulic cylinders widely apply in building, digging, hoisting-and-transport, road machines, automobiles, and also in the processing equipment - metal-cutting machine tools, forge - pressing machines.

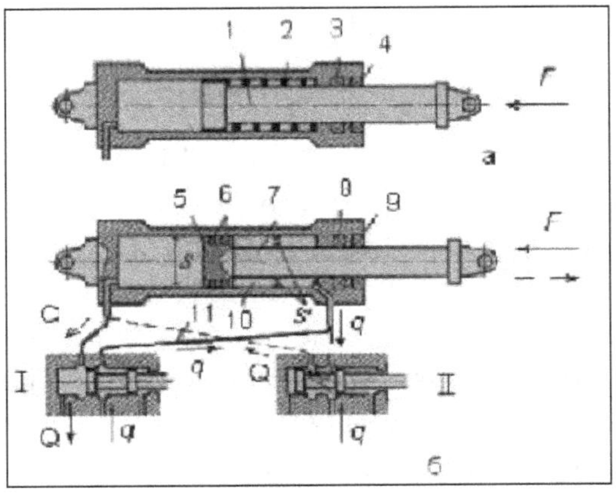

The hydraulic cylinder.

The hydraulic cylinder of one- sided action has a plunger 1, moved by force of pressure of liquid to one side. Reverse motion of the plunger is made under action of external force F if it operates continuously, or spring 2. The unique external condensation of the plunger consists of the basic 3 and antisplash 4 condensing elements. The hydraulic cylinder of double-sided action has the piston 5 with a rod 7, condensed by internal 6 and external 8 sealants. The difference between full S and ring S? areas of the piston conducts to distinction in using pressure p at moving to the left and to the right if overcoming external force F is constant. If constant consumption Q is brought to the cylinder, the difference of the areas results in distinction of speeds of movement of the piston depending on a direction of moving.

In order to eliminate these effects when they are undesirable, such hydraulic cylinders are included by means of a valve under the differential circuit (position I and II of the valve) at which the rod's cavity 10 is continuously connected to a supply circuit 11. If S' = S/2, at the movement to the right (position I of the valve) and to the left (position II of the valve) speed v = Q/S' and force F will be identical.

To get the full symmetry of forces and speeds we apply hydraulic cylinders with a double- sided rod with one internal 1 and two external condensations 3 and 4. In this case the design with the fixed rod is shorter in 1,5 times, than a design with the fixed hydraulic cylinder and a control valve.

The number of consolidations which are the sources of friction and places of external and internal outflows, defines volumetric and mechanical efficiency of the hydraulic cylinder, and also its reliability. From this point of view the hydraulic cylinder with a double-sided rod has the smaller efficiency in comparison with the considered cylinders under other equal conditions.

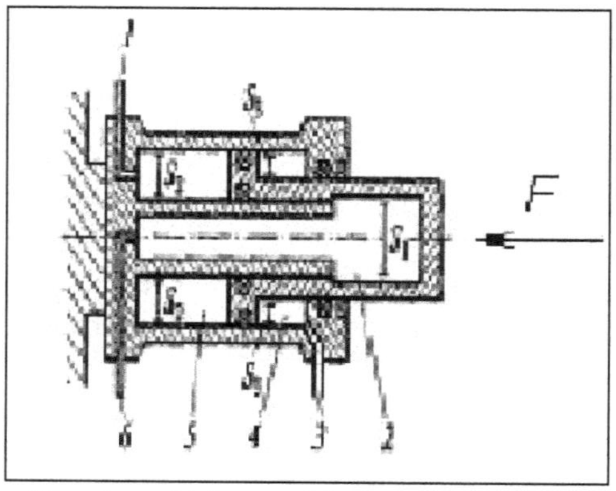

The three-high-speed hydraulic cylinder.

The scheme of the three-high-speed hydraulic cylinder with two levels of developed force is shown in figure. Such hydraulic cylinders are used in the pressing equipment. The fast course of approaching with the speed V1 = Q/S1 is carried out by filling of a cavity through a supply 6 at hydrolines 1 and 3, connected with a plum area. The working course with small speed v2 = Q / (S1+S2) on a short way is carried out at a feeding of cavities 4 and 5 through supplies 1 and 3. Thus the hydraulic cylinder used at the maximum working area, develops the maximum force F = p (S1+S2) at the least pressure p. The fast returnable course with the speed v3 = Q/S3 is made at filling a cavity 4 through a line 3, thus lines 1 and 6 are connected to a plum area.

In many cases hydraulic cylinders work in severe conditions when suddenly changing load- ings and adverse climatic conditions. For protection against the hit of moisture and dirt provide double external consolidations (for example, 3 and 4) antisplash consolidations 2 and 5, and sometimes rubber sylphon 4, entirely closing a rod at promotion. Ending brake devices are established for protection against impacts of the piston against a cover of the hydraulic cylinder at the end of a course. Sockets 2 and 3 are realized in covers of the hydraulic cylinder, and corresponding to them there are cylindrical ledges 7 and 6 on the piston, forming with jacks small backlashes. At the end of a course the ring volume 8 will be squeezed out through a throttle 1 and a ring crack, whose resistance is great, and speed of the piston will decrease. Return valves 5 are stipulated for fast filling of the cylinder in the beginning of a course around of a throttle 1 and of the socket.

Telescopic hydraulic cylinders are applied in the cases when the desirable course exceeds the allowable adjusting length of the hydraulic cylinder. Promotion of the sections of the hydraulic cylinder if it feeds through the line 7 from a source of constant charge Q (for example, the volumetric pump), will occur at different speed value and if overcoming force F is constant, at different pressures.

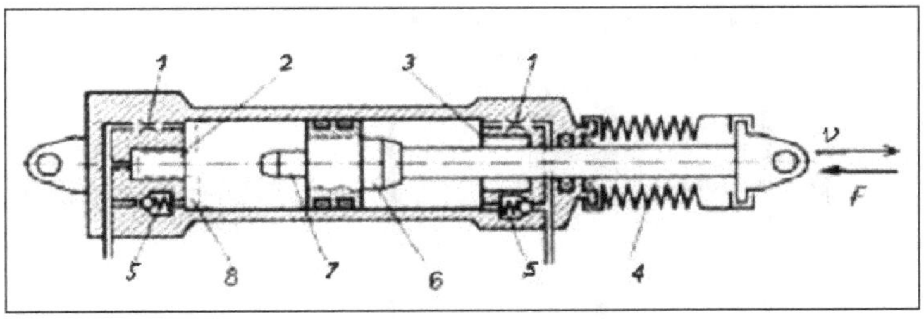

The hydraulic cylinder with trailer throttle brakes and the protected rod.

At promotion the piston 1 displaces the first against the stop with the small speed $V1 = Q/S1$ at the smaller pressure $p1 = F/S1$. After the full promotion of the piston 1 the piston 6, whose area S2 starts to move till full promotion. At the same time, the speed increases up to $V2 = Q/S2$ and the pressure grows up to $p2=F/S2$. The promotion of sections is made under action of force F, or by submission of charge Q through the line 5 in the cavities 2 and 3 through the sleeve 4.

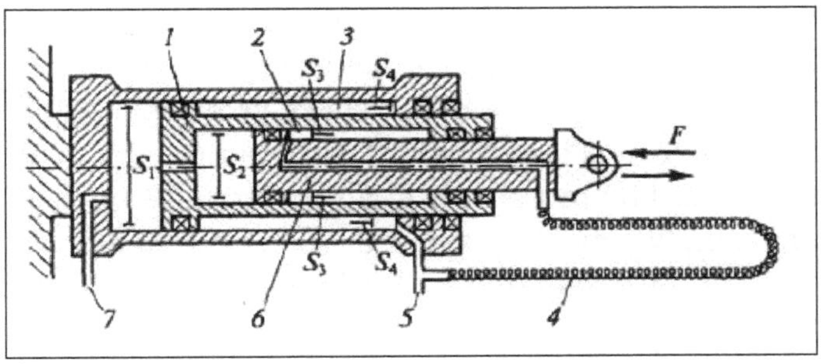

The telescopic hydraulic cylinder.

Volumetric Hydraulic Actuator

Volumetric hydraulic actuator is the set of volumetric hydromachines, hydroequipments, hydrolines (pipelines) and the auxiliary devices, intended to transfer energy and transformation of movement by means of liquid.

There are several types of pumps and hydraulic engines, which can be considered as hydromashines. The hydroequipment are devices which control a hydraulic actuator. These devices adjust the hydraulic actuator and protect it from excessively high and low pressure of liquid. Throttles, flaps of different assignment and hydraulic control valves are concerned to the hydroequipment - devices for changing direction of a liquid stream. As auxiliary devices the so-called conditioners of working liquid providing its quality and condition serve: various separators of firm particles, including filters, heat-exchangers (heaters and coolers of liquid), hydrotanks, and also hydroaccumulators. The listed elements are connected with each other by hydrolines where working liquid goes.

The principle of action of a volumetric hydrodrive is based on small compressibility of drop liquid and transfer of pressure to them under Pascal's law. Pascal's law can be formulated as follows: any change of pressure in any point of the based liquid, without breaking its balance, is transferred in its other points without change. Pascal's law specifies ability of liquid to transfer effort to distance that is widely used in technics(technical equipment). Work of hydraulic jacks, presses, multiplies, brakes, a volumetric hydrodrive, systems hydropneumoautomatics, etc. is based on Pascal's law.

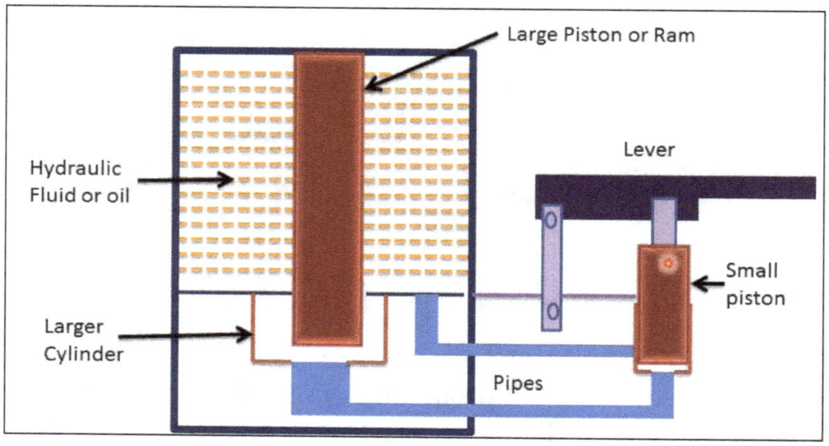

Hydraulic Jack

In this system, a reservoir and a system of valves has been added to a simple hydraulic lever to stroke a small cylinder or pump continuously and raise a large piston or an actuator a notch with each stroke. The top view of within the diagram below shows an intake stroke. An outlet check valve closes by pressure under a load, and an inlet check valve opens so that liquid from the reservoir fills the pumping chamber. The lower view of the diagram shows the pump stroking downward. An inlet check valve closes by pressure and an outlet valve opens. More liquid is pumped under a large piston to raise it. To lower a load, a third valve (needle valve) opens, which opens an area under a large piston to the reservoir. The load then pushes the piston down and forces the liquid into the reservoir.

Stabilization and Synchronization of Movement of Target Parts

In hydraulic actuators of metal-cutting machine tools and other machines it is frequently applied throttle regulation with the device for stabilization of speed of an output link with the purpose of improvement of loading characteristics. The regulator of the stream that provides stability of established speed irrespective from the change of loading serves as such device.

The scheme of a hydraulic actuator of linear motion with the regulator of a stream established consistently on an output from the hydraulic cylinder is shown in figure. The regulator 7 consists of an adjustable throttle 2 and a reducing valve 3. The last supports the constant pressure Pth upon the input in a throttle. Pressure is belived to be constant and equal to atmospheric on the output from the throttle at small resistance of a removing hydroline.

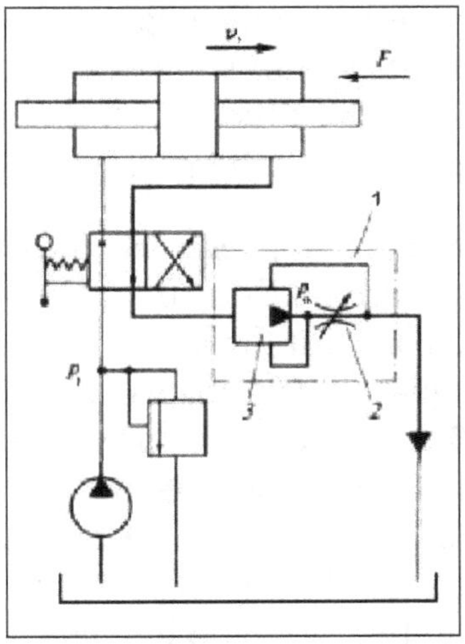

The scheme of a hydraulic actuator with a regulator of a stream.

Hence, the charge of liquid through a throttle will be steady. Submission of liquid in hydraulic engine Qhe = Qp - Qth is steady when the submission of the pump is steady and does not depend on loading, so speed of an output link will also be steady. Actually speed Vp with increase in loading decreases owing to outflow in the pump, growing with increase in pressure, and also due to discrepancy of work of a reducing valve.

Hydraulic actuator with a steady output frequency of rotation when frequency of rotation of a shaft of the pump is variable and when loading on a shaft of the hydraulic motor is variable is called stabilized.

Stabilization of output frequency of rotation is usually carried out on a signal of the centrifugal or electrohydraulic regulator influencing through an intensifying part on the mechanism of change of working volume of the pump. The hydromotor in such circuit usually uncontrollable.

The scheme of the elementary hydraulic actuator for stabilization of rotation frequencies of the alternating current stable frequencies electrogenerator with a centrifugal regulator of speed is shown in figure. The hydraulic actuator consists of the pump 1 connected to the reducing engine, and the hydromotor 2, the shaft 3 which is connected to the electrogenerator and results in rotation of a centrifugal regulator 4 which moves the valve 5 of hydraulic control valves of a control system with changing of rotation frequency of the hydromotor 2. The last, submitting liquid into the hydraulic cylinder 6 of the mechanism of regulation (the inclination angle of the disk) the pump 7, restores mismatch, supporting output rotation frequency of the hydromotor at constant value.

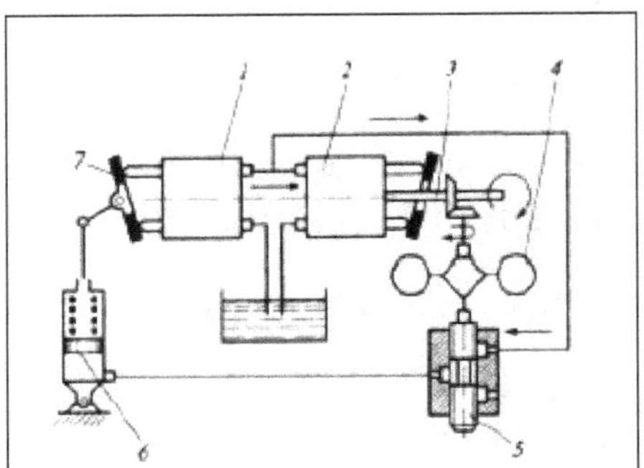

The scheme of a hydraulic actuator with steady output rotation frequency.

When one pump actuates two or more hydraulic engines, sometimes it is necessary to provide identical speed of their target parts, despite of some distinction of loadings. Such requirement arises, for example, in a hydraulic actuator of the load-lifting machine in which the cargo is risen up by several parallel-connected hydraulic cylinders without skews.

Synchronization of movement of hydraulic engines is carried out by means of a so-called divider of a stream which principle of action is based on throttling. At point M the stream branches into two ones, each of which passes through a constant throttle 7, and then is brought to a sleeve 5 with the floating piston 4. The floating piston plays a role of a valve, moving in one or other side depending on difference of working pressure. The difference of pressure arises when the charge of liquid in one branch is distinct from the charge in another one owing to different loadings and, hence, loss of pressure in one throttle more, than in the other one. Moving aside smaller pressure, for example to the right, the piston 4 reduces the area of an aperture 2 and increases the area of an aperture 3. The piston will stop when pressure in the right and left cavities of a sleeve and consequently, the charge through these cavities will be equal.

Advantages and Disadvantages of Hydraulic Actuators

Variable hydraulic actuators are widely used as drives of machine tools, rolling mills, pressing and the foundry equipment, road and building machines, transport and agricultural machines, etc. A number of advantages in comparison with mechanical and electric transfers explains such their wide application:

- Infinitely variable control of gear-ratio in a wide range and an opportunity to create the big reduction ratio;
- Small specific weight, i.e. the weight of a hydroactuator is in ratio to transmitted capacity (0,2,0,3 kg / kWt);
- Opportunity of simple and reliable protection of the engine from overloads;
- Small sluggishness of the rotating parts, providing fast change of operating modes (start-up, dispersal, a reverser, a stop);
- Simplicity of transformation of rotary movement into reciprocating one;
- Opportunity of positioning a hydraulic engine on removal(distance) from an energy source and freedom in making configuration.

It is also necessary to reckon with disadvantages of hydraulic actuators:

- Efficiency of a volumetric hydraulic actuator is a little bit lower, than efficiency of mechanical and electric transfers, and during regulation it is reduced;
- Conditions of operation of a hydraulic actuator (temperature) influence its characteristics;
- Efficiency of a hydraulic actuator is a little reduced in the process of exhaustion of its resource owing to the increase in backlashes and the increase of outflow of liquid (falling of volumetric efficiency);

- Sensitivity to pollution of working liquid and necessity of high culture service.

Application Areas of Pneumatic and Hydraulic Actuators

Pneumatic Processing Centers

The scheme of the pneumatic processing center with a rotary table is shown in figure below.

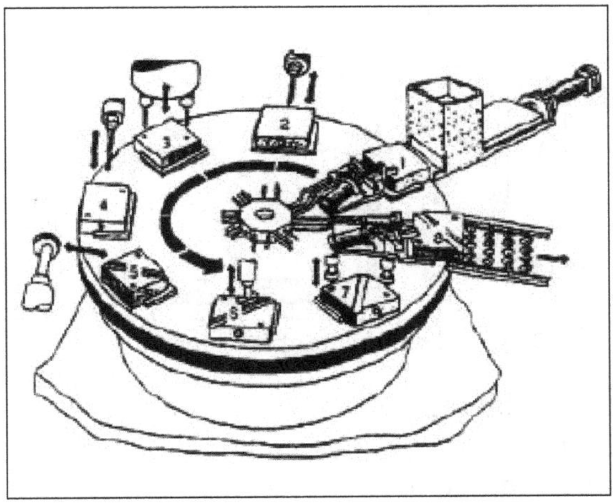

The scheme of the pneumatic machining center.

The pneumatic processing center carries out the following functions: - step-by-step rotation with the help of a rotary table - parts cutting with the help of tools 2-7 (drilling, grinding, milling); - reception, positioning and fixing a detail with the help of the block 1; - delivery of a detail with the help of the block 8. On the rotary table various processes of serial part cutting can be carried out. Thus, only once the detail goes into the center and is established for processing and only once removed and given out from the pneumatic processing center center.

The scheme of the pneumatic processing center for sawing material's is shown in figure below.

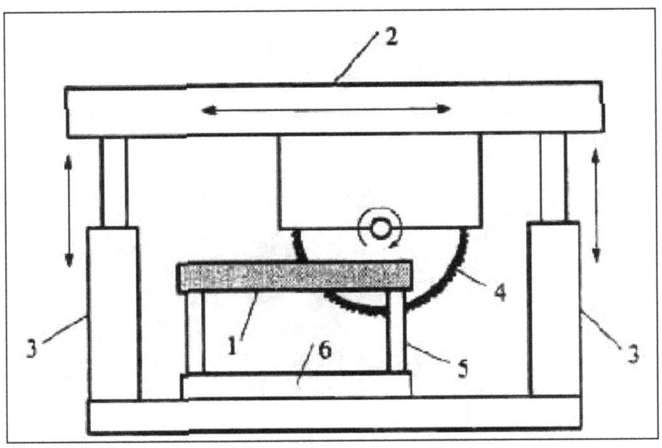

The scheme of the pneumatic processing center
for sawing materials.

All actuators of the center, including a drive of the tool are pneumatic. Simultaneous movement of several drives achieves any trajectory of sawing material.

Batching

Batching of liquids and the fused metals, for example, in foundry manufacture can be made by system which circuit is shown in figure below.

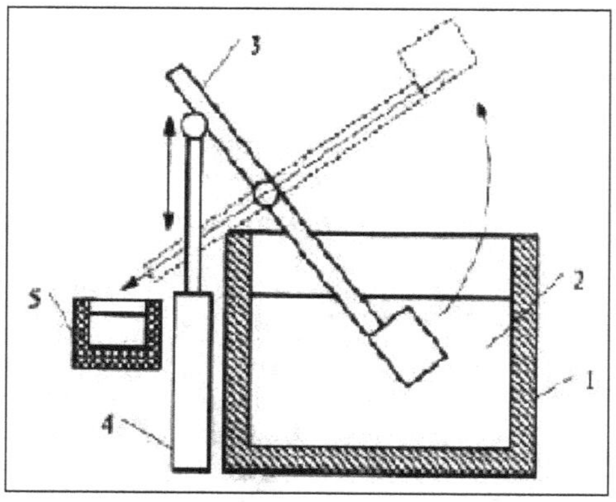

The scheme of batching.

The lever with a ladle is actuated by the power cylinder. The volume of a ladle corresponds to a required doze of liquid which gets in the top position of a rod of the power cylinder. With the rod moving to the bottom position the liquid from the ladle is automatically poured in accepting chamber.

Robotics

On the basis of pneumatic actuators stationary and mobile robots with wide functionalities can be created.

One of the first stationary Russian serial pneumatic industrial robots is MÏ - 9Ñ The robot has four degrees of mobility provided by horizontal and vertical power cylinders, and also by a turning drive of gripping device and a drive of fingers.

Mobile pneumatic robots allow to realize walking type of moving. The example scheme of such robot for automation of mine clearing is shown in figure below.

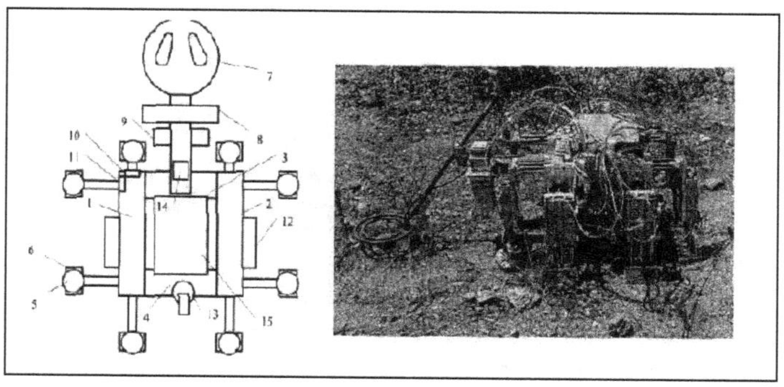

The scheme of the mobile robot.

The transport module consists of longitudinal and cross-section pneumatic cylinders which cases are connected symmetrically. Each transport cylinder has two manipulators which are established at end-walls of rods of the cylinder.

Manipulator consists of the elevating cylinder with a course, allowing to overcome the set sizes of obstacles, and supports with a gear working surface to improve coupling with a surface when it makes along on slopes. The block of mines detection consists of metal detector, the infrared sensor and a chemical sensor control. This block is established on a forward part on a course of the robot. The sensor of longitudinal position movement is placed on the case of the longitudinal cylinder. The sensor of cross-section position movement and the block of detection is, accordingly, on the case of the cross-section cylinder.

The block of valves settles down from both sides of the robot that provides the minimal length of onboard feeding lines. They submit compressed air from the site of the feed component connected to the main feeding line on to managing valves.

The feed component provides robot rotation concerning the main feeding line without creation of the counter torque at the expense of sliding pneumatic condensation. The onboard computer and an electronic compass are established on the top part of the platform.

The robot has two control modes. The first one is a transport mode. In this case longitudinal cylinders move the manipulators which are in the top position, at the maximum speed, and full length of a working course.

During this movement the robot leans on the moving surface by means of manipulators of the cross-section cylinder. On the following step manipulators of longitudinal cylinders carries out this role and so on. The robot has an opportunity to change a direction of movement on a required angle. Simultaneous movement of longitudinal cylinders in opposite directions carries out rotation of the robot when their manipulators contact with the surface of movement. The second control mode is the mode of mines detection. In this mode the block of detection scans a zone in front of the robot. The trajectory of scanning is realized by cross-section and longitudinal cylinders, which are actuated with the rated speed corresponding to parameters of the block of detection.

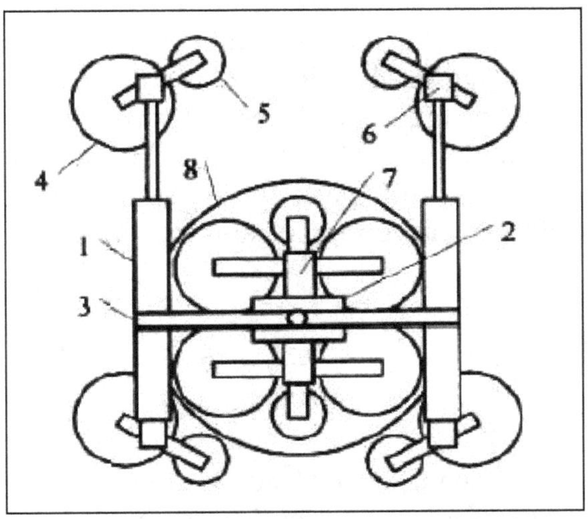

The scheme of the mobile robot with vertical displacement.

Creation of robots of vertical moving is rather new direction in the development of mobile robots. These robots are capable to move and carry out technological operations on surfaces inclined at any corner of an that determines the efficiency of their use at big heights in construction, in radioactive zones at nuclear reactors service and in other extreme situations.

The robot consists of the longitudinal movement module and the rotating movement module. These modules are connected with each other along its rotation axis. The longitudinal movement module consists of two pneumatic cylinders.

Manipulators of longitudinal movement are established at end-walls of cylinders rods. They have vacuum gripping devices which can move concerning the case of the robot by using elevating cylinders.

The rotation module contains the pneumatic drive of rotation connected rigidly with ma- nipulators, having the same combination of gripping devices and elevating cylinders, as the longitudinal movement module.

The robot has nine degrees of freedom. Movements of the robot are carried out by using step-by-step vacuum fixing of the lon- gitudinal movement module and the rotation module to a surface of movement and serial moving of cylinder rods and their cases to a direction of movement with an opportunity of turn with the help of a drive of rotation on a required angle. Longitudinal movement and rotary movement modules are equipped with elevating cylinders 5. Elevating cylinder rods are connected with concentric gripping devices and sealing gripping devices. This connection is elastic in the direction of the cylinders action. Such connection enables all working surface of gripping devices to enter contact with a rough surface of movement.

Carbon Nanotube Actuators

The exceptional electrical and mechanical properties of carbon nanotubes have made them alternatives to the traditional electrical actuators for both microscopic and macroscopic applications. Carbon nanotubes are very good conductors of both electricity and heat, and are also very strong and elastic molecules in certain directions. These properties are difficult to find in the same material and very needed for high performance actuators. For current carbon nanotube actuators, multi-walled carbon nanotubes (MWNTs) and bundles of MWNTs have been widely used mostly due to the easiness of handling and robustness. Solution dispersed thick films and highly ordered transparent films of carbon nanotubes have been used for the macroscopic applications.

Microscopic Applications

Carbon Nano-tweezers

Carbon nanotube tweezers have been fabricated by deposition of MWNT bundles on isolated electrodes deposited on tempered glass micropipettes. Those nanotube bundles can be mechanically manipulated by electricity and can be used to manipulate and transfer micro- and nano-structures. The nanotube bundles used for tweezers are about 50 nm in diameter and 2 μm in lengths. Under electric bias, two close sets of bundles are attracted and can be used as nanoscale tweezers.

Nanotube on/off Switches and Random Access Memory

Harvard researchers have used the electrostatic attraction principle to design on/off switches for their proposed nanotube Random Access Memory devices. They used carbon nanotube bundles of ≈50 nm in diameter to fabricate their proof-of-concept prototypes. One set of MWNT bundles are laid on the substrate and another set of bundles is trenched on top of the underlying nanotube bundles with an air gap in between them. Once electrical bias is applied, the sets of nanotube bundles are attracted, thus changing the electrical resistance. These two states of resistance are on and off states. Using this approach, more than 10 times the difference between off and on state resistances has been achieved. Furthermore, this idea can be used to create very highly packed arrays of nanoswitches and random access memory devices, if they can be applied to arrays of single-walled carbon nanotubes, which are about 1 nm in diameter and hundreds of micrometres in length. The current technical challenge with this design is the lack of control to place arrays of carbon nanotubes on substrate. This method is followed by some researches at Shahid Chamran University of Ahvaz as well.

Carbon Nano-heat Engine

A research group at Shanghai University led by Tienchong Chang have found a dominoe-like motion in carbon nanotubes, which can be reversed by translating direction when different temperatures are applied. This phenomenon makes it possible to use carbon nanotubes as a heat engine working between two heat sources.

Macroscopic Applications

Nanotube Sheet Electrodes as Actuators

Researchers of AlliedSignal initially demonstrated the possibility of electrically powered actuators fabricated by carbon nanotube sheets. They taped carbon nanotube sheets on two sides of a double sided scotch tape and applied potential on the nanotube sheets in a NaCl electrolyte solution. Nanotube sheets are used as electrolyte-filled electrodes of a super capacitor. Nanotube sheets are electrically charged by the double layer formation at the nanotube-electrolyte interface without any need of ion intercalation. Therefore, electrically driven actuators of nanotube sheets are superior to the conjugated polymer actuators which involve solid-state dopant diffusion and structural changes limiting rate, cycle life and energy conversion efficiencies. On the other hand, ferroelectric and electrostrictive materials are also very useful for direct energy conversion, but they require high operation voltages and ambient temperature of a limited range. Nanotube sheet actuators were shown to operate at low voltages (≈1 Volts or less) and provide higher work densities per cycle than other alternative technologies. Later Baughman et al. showed that actuator response can be observed up to switching rates of 1 kHz and cycling the nanotube actuator at constant rate of 1 Hz for 140000 cycles decreases the stroke by ≈33%. 0.75 MPa of stress were measured on the nanotube sheet actuators, which is greater than the maximum stress (0.3 MPa) that can be loaded on a human muscle.

The maximum actuator strain for electrically driven actuators of carbon nanotube sheets can be improved up to 0.7% in a 1 M electrolyte once the sheets are annealed in an inert atmosphere at very high temperatures (1100 °C) in contrast to once reported 0.1% or less for low electrochemical potentials (≈1 V or less). The maximum strain for the carbon nanotube sheet actuators at low voltages is greater than that of the high modulus ferroelectric ceramic actuators (≈0.1%), but it is lower than that of the low voltage (≈0.4 V) conducting polymer actuators (≈3% film direction, 20% thickness direction). Strokes were reported as high as 215% for strain biased low modulus electrostrictive rubbers under biases greater than 1kV (corresponding to an electric field 239 MV/m for the geometry). Spinks et al. realized pneumatic actuation from the carbon nanotube sheets in electrolyte solutions with high electrochemical potential (1.5 V), which cause gas generation in the electrolyte. The released gas dramatically increases the actuator stroke from the carbon nanotube sheet. Thickness of the carbon nanotube sheet expands by ≈300% and the sheet plane contracts by 3%.

Artificial Muscles and Giant Strokes by MWNT Aerogel Sheets

Highly ordered free standing aerogel sheets of MWNTs can be realized by simply drawing the sheet from the sidewalls of CVD grown MWNT forests. UT Dallas researchers came up with the conventional method where they attach an adhesive tape to the sidewalls of MWNT forests and they pull the tape at a constant rate as fast as 7 meters per minute to get 3–5 cm wide aerogel sheets of aligned MWNTs which have exceptional mechanical and optical properties. The aerogel sheets have a density of ≈1.5 mg/cm^3, an areal density of 1-3 μg/cm^2 and a thickness of ≈20 μm. The thickness is decreased to ≈50 nm by liquid-based densification to decrease the volume. The aerogel sheets can be stretched as much as three times along the width while low-modulus rubber like behavior is remained.

Having aerogel sheets of MWNTs, UT researchers fabricated actuators with giant strokes (≈180% actuation along the width) with 5 ms delay time between applying the potential and observing the maximum stroke. Therefore, the actuation rate is slightly better than that of the human muscle. This is a very important achievement considering the actuation rate for artificial muscles used in robots is typically much slower. Furthermore, the use of carbon nanotubes as the building blocks as an artificial muscle also helps in terms of strength and robustness by making the artificial muscle stronger than steel in one direction and more flexible than rubber in the other two directions. The lack of electrolyte solution and temperature robustness of the aerogel sheet in inert ambient makes high temperature operation possible. The actuation stroke decreases by only 50% from its room temperature value to 1344 °C. Thus, this design of artificial muscles can be quite useful for many industrial applications with the drawback of high voltage operation for giant strokes.

Challenges and Future Applications

As a result, carbon nanotubes have been shown to be great materials for actuation related applications. The subfield of carbon nanotube actuators have been quite successful and ready for scalable applications considering there are quite a few conventional and scalable methods for the synthesis of large scale carbon nanotubes. Carbon nanotube sheets used as electrodes in electrolyte solutions offered low voltage operations at room temperature with actuation strokes and rates comparable to the conducting polymer actuators, but with higher work densities per cycle and life times. However the actuation strokes are much smaller than those of the electrostrictive rubbers which operate at three orders of magnitude higher voltages. On the other hand, realization of carbon nanotube aerogels made giant strokes possible comparable to electrostrictive rubbers at room temperature, but carbon nanotube aerogels can perform at a very wide range of temperatures, and with very high actuation rates, which are even better than the actuation rate of the human muscles.

Piezoelectric Actuator

A piezoelectric actuator converts an electrical signal into a precisely controlled physical displacement (stroke). If displacement is prevented, a useable force (blocking force) will develop. The precise movement control afforded by piezoelectric actuators is used to finely adjust machining tools, lenses, mirrors, or other equipment. A piezo actuator can be used to control hydraulic valves, act as a small-volume pump or special-purpose motor, and in other applications requiring movement or force.

Two different types of piezo actuators are stack actuators and stripe actuators.

1. Stack Actuator: Piezo stack actuators offer low stroke and a high blocking force. Based upon the user's requirements stack actuators can be either discrete or co-fired.

Discrete stacks (high-voltage stack actuators) are composite structures made by stacking separately finished piezoelectric ceramic discs or rings and metal electrode foils with an adhesive. Operating voltages ranging from 500 V thru 1,000 V are typical.

Co-fired multilayer stack actuators, also called "monolithic stacks", involve no adhesive, but rather a high temperature sintering of the complete ceramic and electrode pile. Operating voltages of a co-fired stack are up to 200 V. Rectangular cross sections are typical due to the ease of cutting processes in production.

Both the discrete and co-fired stacks can be insulated with a coating material — a bare stack — or encased in stainless steel when protection from mechanical stress and environmental extremes are needed.

2. Stripe Actuator: A Stripe actuator, also called a bending actuator, is designed to produce a

relatively large mechanical deflection in response to an electrical signal. This deflection offers a large stroke and a very limited blocking force when compared to a stack actuator.

In a stripe actuator, two thin layers of piezoelectric ceramic are bonded together, usually with the direction of polarization coinciding, and are electrically connected in parallel. When electrical input is applied, one ceramic layer expands and the other contracts causing the actuator to flex.

Benefits of Piezo Actuators

Piezoelectric actuators have the following attributes:

- They have a theoretical unlimited resolution, often resolving to sub-nanometer values, i.e. smallest changes in supply voltage are detected and converted into linear motion without skipping or stepping;
- High actuation forces can be attained without significant loss in precision, for instance loads higher than 10.000 Newtons can be positioned with micrometer accuracy;
- Extremely high response times of under 1 millisecond. As an example, the elongation response of a piezoelectric actuator can be limited only by the speed at which sound is propagated through ceramic materials. Accelerations several thousand times higher than gravitational acceleration (G) can be attained;
- There are no moving parts and thus no friction or free play. Elongation of a piezoelectric actuator is based only on deformation of the material and there is also no fatigue or aging involved. Endurance tests have proved that there is no altering in such an actuator's operation even after 500 million cycles;
- Very low power consumption as the piezo effect converts energy directly into motion, the execution element absorbs energy only when elongating;
- No magnetic fields are generated when the actuator is operating.

Such advantages have made piezo actuators very popular in several domains as:

- Precision mechanics and Mechanical engineering – Found in adjust and extrusion tools, correction of wear and active control in tools injection nozzle control, micropumps, piezo hammers, micro etching, active vibration cancelling systems, micro and nanorobots;
- Optics and measurement systems – Employed in rapid scanning and positioning of mirrors, holography, interferometry, laser sweeping, fiber optics positioning, image stabilization, active adaptive auto focus;
- Medicine – Micromanipulation and microsurgery, cell penetration, microdosing systems, physiological stimulations, shock generation;
- Microelectronics – Mask positioning, microlithography, inspection and control systems.

Principles of Operation

A piezoelectric actuator is an electrically controlled positioning element that functions based on the piezoelectric effect. The direct piezo effect, employed for instance in piezoelectric force sensors, represents generating an electric charge as an effect of mechanical strain. Actuators are based on the reverse effect, namely and electrical field parallel to the direction of polarization, determines an elongation of the crystalline material with respect to the same direction. The electrical field generates a torque over the electrical dipoles found in the structure of the material, which will be aligned along the field, producing in turn a change in the length of monocrystalline partitions.

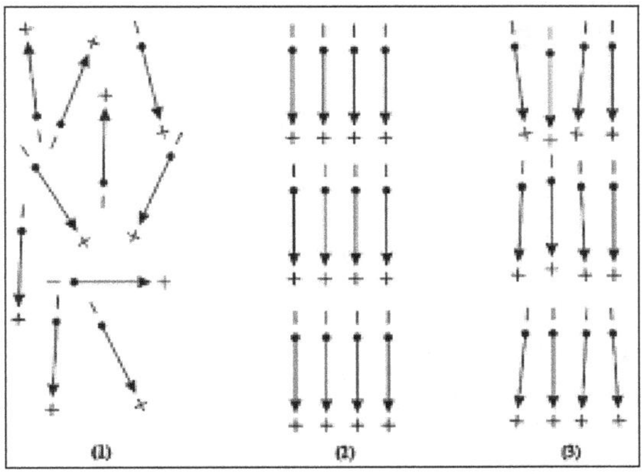

Dipoles aligning in an electrical field.

However in natural monocrystalline materials, such as quartz, tourmaline, Rochelle salt, and so on, the piezoelectric effect has small ampleness so that synthetic polycristalline ceramic materials with superior piezoelectric have been developed, such a material being lead zirconate titanate (PZT). These materials are produced in a wide range of variants and are widely employed in piezoelectric actuators manufactured today.

A piezoelectric transducer is an actuator that elongates when subjected to an electrical field. This elongation is dependent to the ceramic material employed, its length, intensity of the electric field and force acting upon it. An expression to summarize this is:

$$\Delta L = E \cdot d_{ij} . L_O + \frac{F}{c_T} = \Delta L_O + \frac{F}{c_T}$$

where,

ΔL - Elongation (meters)

L - Transducer length (meters)

E - Electric field intensity (V/m)

F - Force acting upon the transducer (N)

d_{ij} - Constmt of material (m / V)

c_T - Elastic cons tan t of material (N/m)

Maximum elongation is limited by the intensity of the electric field at which the actuator can be subjected without being destroyed by tension shocks, limits usually around 1-2 kV/mm. A transducer is an elastic body with a certain rigidity, same as any other solid. If compressed with a certain force F, it will be compressed with the amount of force divided by the specific constant for the material.

Two types of behavior of a piezo transducer can be identified, based on the type of strain it is subjected at:

- Constant load (F constant) – In this case the force is usually exerted by a weight that remains constant during the elongation process of the actuator. Since elongation capabilities are not affected by load, maximum elongation will be the same as when the material is not mechanically stressed, but the zero point of the positioning domain will be displaced down with the value F/constant of material.

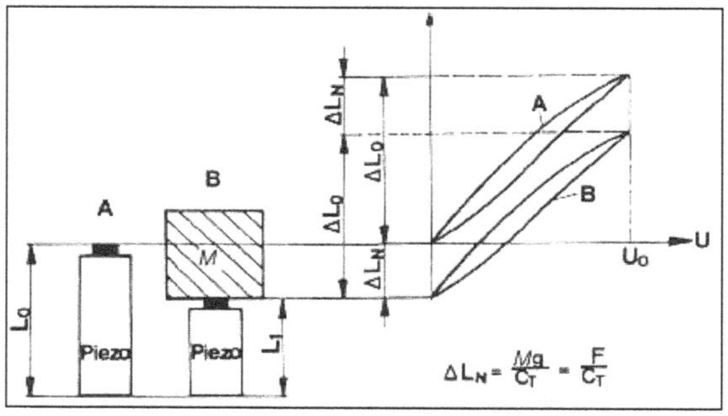

Piezo transducer under constant load.

- Variable load (F as a function of elongation) – Here force is dependent to the elongation of the piezo material. All application in which a positioning element acts upon a spring or an elastic wall find themselves in this category. The more rigid the elastic obstacle – spring or elastic wall – with higher elastic constant Cs, the less travel is available to the transducer.

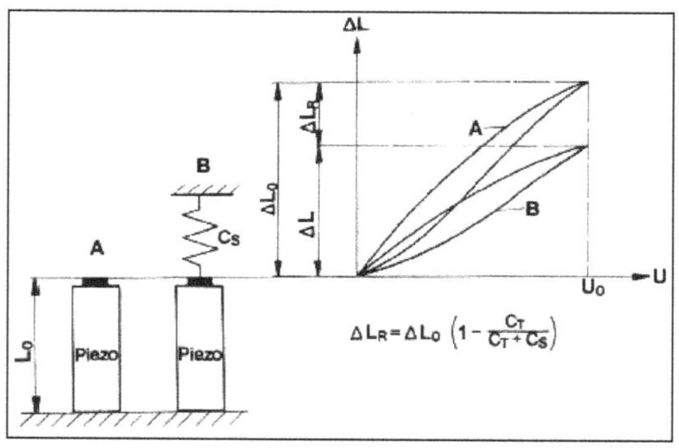

Piezo transducer under variable load.

Actual elongation will be:

$$\Delta L = \Delta L_O \cdot \frac{C_T}{C_T + C_S} < \Delta L_O$$

As it can be seen in the graphical representations above, the elongation of a transducer is not linearly variable with supply voltage, but rather a hysteresis curve, with different values as voltage rises or decreases. The hysteresis phenomena is determined by crystalline polarization effects – as seen in figure - and molecular effects, maximum values of this effect are around 10 to 15 percent of the desired travel. These values translate into errors in the micrometer domain, and can be insignificant in certain applications. If more precision is required a closed-loop control of elongation can be implemented, by means of a position reactive system.

Dynamic Behavior

A piezo actuator can reach its designed elongation in a time equal to one third of its resonant frequency cycle. The resonant frequency of an unstressed transducer is of several tens of kHz for travel in the range of micrometers and several kHz for travel higher than 100 mm. A supplementary load will lower the resonant frequency according to the expression:

$$f_o = \frac{1}{2\pi} \cdot \sqrt{\frac{c_T}{m_{eff}}}$$

The actual value of mass can be inferred from the representation below:

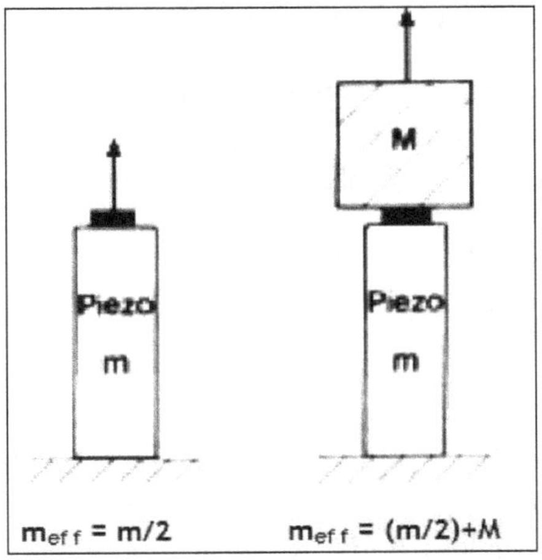

Helical Band Actuator

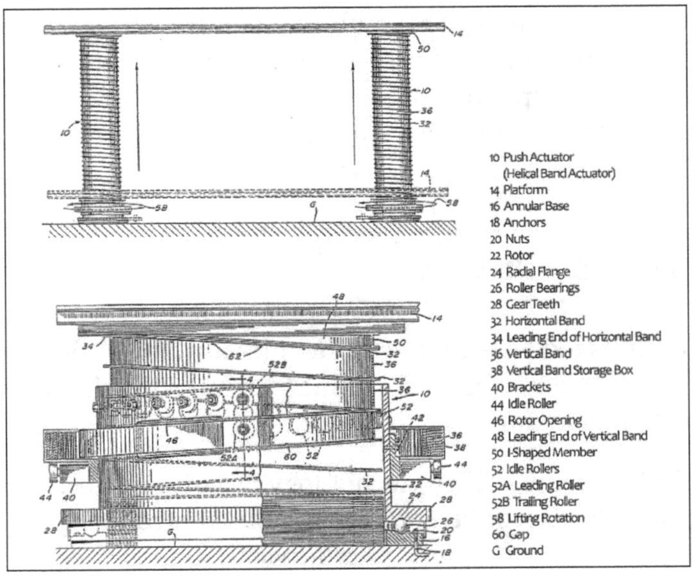

Patent Drawing for Spiralift Actuator.

A helical band actuator, generally known by the trademark Spiralift, is a complex and specialized linear actuator used in stage lifts and material handling lifts. The actuator forms a high-capacity telescoping tubular column (lifting capacities to 25,000 pounds, travel to 40 feet).

Pierre Gagnon was awarded a US patent for the "Push Actuator" in 1989. Gagnon developed the actuator to substitute hydraulic cylinders in stage and orchestra lift systems. Hydraulic jacks were the predominant push actuators used in performing arts facilities, but they had their issues.

The piston rods of a hydraulic lift extend below the platform they support by more than the length that the lift travels. For lifts at grade, housings for the rods are often sunk into support holes, known as caissons, that are inaccessible and often complicated by groundwater levels. If placed above ground, the space occupied by the rod housings is unusable. Gagnon's actuator eliminates caissons and reduces the volume of unusable space below the platform of the lift.

Another significant issue with hydraulic lifts is platform drift, the tendency of a platform to lower due to depressurization of the hydraulic system over time. Helical band actuators do not suffer from platform drift because they are rigid mechanical devices.

Traveling nut screw jacks and rigid chain actuators can also be effective at addressing these issues with hydraulic jacks, and are also used for stage lifts.

Principle of Operation

The telescoping column is formed by a pair of interlocking stainless steel bands. One band has a vertical rectangular profile and the other horizontal, much like an oversized Slinky. The vertical band spirals up on itself into a stacked helix, forming the wall of the column, while at the same time, the horizontal band interlocks the continuous spiral seam of the vertical band. When the column lowers, the bands separate and retract into two compact coils. The bands are combined, separated and stored by an assembly located at the base of the column. The result is an efficient (50%-80%) telescoping lifting column.

To incorporate the device into a lift system, multiple helical band actuators are arranged below the lift platform where they are powered by an electric motor(s) and synchronized transmission. Helical band actuators require a separate lateral support mechanism, usually provided in the form of guide rails or self-guiding frame, such as a pantograph (e.g. a scissor lift).

MEMS Magnetic Actuator

A MEMS magnetic actuator is a device that uses the microelectromechanical systems (MEMS) to convert an electric current into a mechanical output by employing the well-known Lorentz Force Equation or the theory of Magnetism.

Micro-Electro-Mechanical System (MEMS) technology is a process technology in which mechanical and electro-mechanical devices or structures are constructed using special micro-fabrication techniques. These techniques include: bulk micro-machining, surface micro-machining, LIGA, wafer bonding, etc.

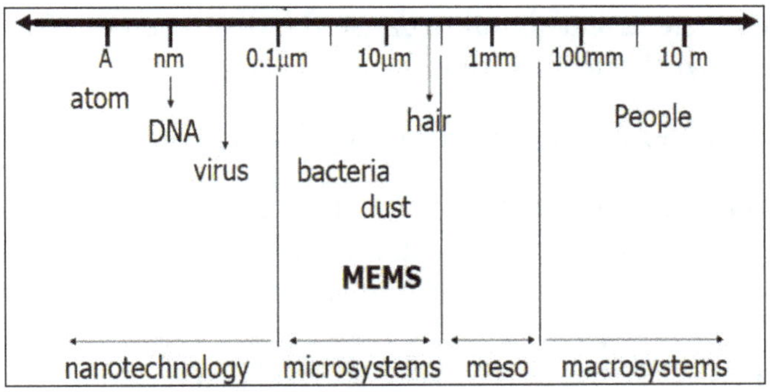

MEMS scale of dimensions.

A device is considered to be a MEMS device if it satisfies the following:

- If its feature size is between 0.1 μm and hundreds of micrometers. (below this range, it becomes a nano device and above the range, it is considered a mesosystem).
- If it has some electrical functionality in its operation. This could include the generation of voltage by electromagnetic induction, by changing the gap between 2 electrodes or by a piezoelectric material.
- If the device has some mechanical functionality such as the deformation of a beam or diaphragm due to stress or strain.
- If it has a system-like functionality. The device must be integrable to other circuitries to form a system. This would be the interfacing circuitry and packaging for the device to become useful.

For the analysis of every MEMS device, the Lumped assumption is made: that if the size of the device is far less than the characteristic length scale of the phenomenon (wave or diffusion), then there would be no spatial variations across the entire device. Modelling becomes easy under this assumption.

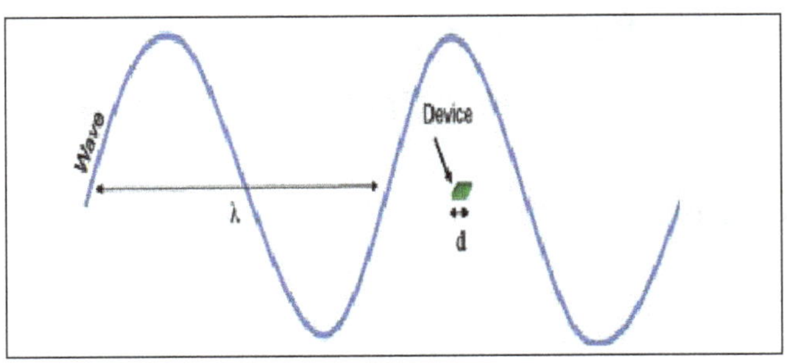

The Lumped Assumption.

Operations in MEMS

The three major operations in MEMS are:

- Sensing: Measuring a mechanical input by converting it to an electrical signal, e.g. a MEMS accelerometer or a pressure sensor (could also measure electrical signals as in the case of current sensors).

- Actuation: Using an electrical signal to cause the displacement (or rotation) of a mechanical structure, e.g. a synthetic jet actuator.
- Power generation: Generates power from a mechanical input, e.g. MEMS energy harvesters.

These three operations require some form of transduction schemes, the most popular ones being: piezoelectric, electrostatic, piezoresistive, electrodynamic, magnetic and magnetostrictive. The MEMS magnetic actuators use the last three schemes for their operation.

Magnetic actuation

The principle of magnetic actuation is based on the Lorentz Force Equation.

$$\vec{F}_{mag} = q\vec{v} \times B$$

When a current-carrying conductor is placed in a static magnetic field, the field produced around the conductor interacts with the static field to produce a force. This force can be used to cause the displacement of a mechanical structure.

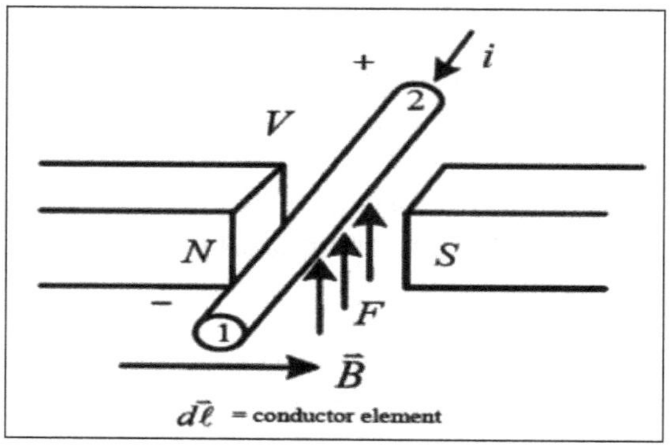

Magnetic actuation

Governing Equations and Parameters

A typical MEMS actuator is shown on the right. For a single turn of circular coil, the equations that govern its operation are:

- The H-field from a circular conductor:

$$H(z) = \frac{Ir^2}{2(r^2 + z^2)^{3/2}}$$

- The force produced by the interaction of the flux densities:

$$F_z = B_I A_{mag} \int_{z}^{z+h_{mag}} \frac{dHz}{dz} dz$$

The deflection of a mechanical structure for actuation depends on certain parameters of the device. For actuation, there has to be an applied force and a restoring force. The applied force is the force represented by the equation above, while the restoring force is fixed by the spring constant of the moving structure.

The applied force depends on both the field from the coils and the magnet. The remanence value of the magnet, its volume and position from the coils all contribute to its effect on the applied Force. Whereas the number of turns of coil, its size (radius) and the amount of current passing through it determines its effect on the Applied Force. The spring constant depends on the Young's Modulus of the moving structure, and its length, width and thickness.

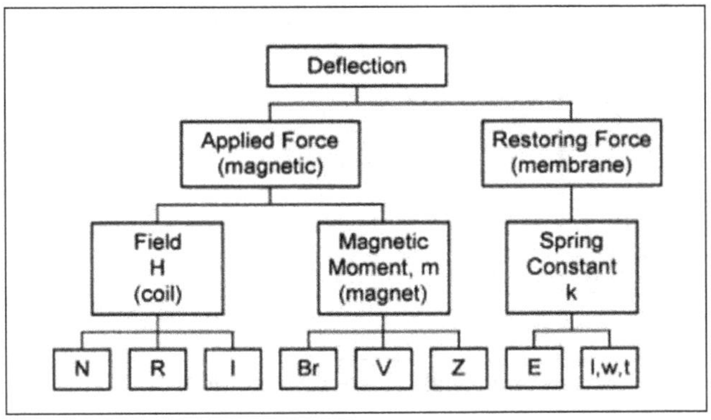

Magnetostrictive Actuators

Magnetic actuation is not limited to the use of Lorentz force to cause a mechanical displacement. Magnetostrictive actuators can also use the theory of magnetism to bring about displacement. Materials that change their shapes when exposed to magnetic fields can now be used to drive high-reliability linear motors and actuators. An example is a nickel rod that tends to deform when it is placed in an external magnetic field. Another example is wrapping a series of electromagnetic induction coils around a metal tube in which a Terfenol-D material is placed. The coils generate a moving magnetic field that courses wavelike down the successive windings along the stator tube. As the traveling magnetic field causes each succeeding cross section of Terfenol-D to elongate, then contract when the field is removed, the rod will actually "crawl" down the stator tube like an inchworm. Repeated propagating waves of magnetic flux will translate the rod down the tube's length, producing a useful stroke and force output. The amount of motion generated by the material is proportional to the magnetic field provided by the coil system, which is a function of the electric current. This type of motive device, which features a single moving part, is called an elastic-wave or peristaltic linear motor.

Advantages of Magnetic Actuators

- High actuation force and stroke (displacement).
- Direct, fully linear transduction (in the case of electrodynamic actuation).
- Bi-directional actuation.
- Contactless remote actuation.
- Low-voltage actuation.

- A figure of merit for actuators is the density of field energy that can be stored in the gap between the rotor and stator. Magnetic actuation has a potentially high energy density.

Magnet Material

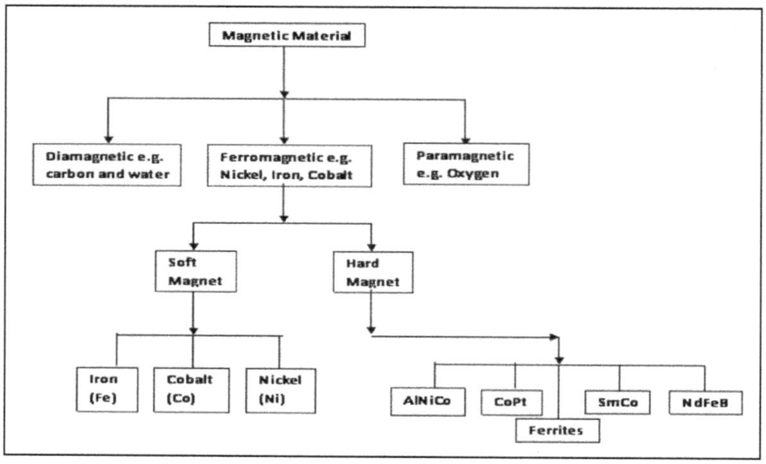

Magnet material selection for static B-Field.

The operation of the magnetic actuator depends on the interaction between the field from an electromagnet and a static field. To produce this static field, it is important to use the right material. In MEMS, permanent magnets have become the favorite because they have a very good scaling factor and they retain their magnetization even when there is no external field meaning that they need not be continuously magnetized when they are in use.

Integrating the Magnet into the MEMS Device

MEMS devices are designed and fabricated using special micro-fabrication techniques. The major challenge however for magnetic MEMS is the integration of the magnet into the MEMS device. Recent research has suggested solutions to this challenge.

Fabrication (or Molding) of the Magnet

There are several ways by which the magnet could be fabricated on a MEMS structure:

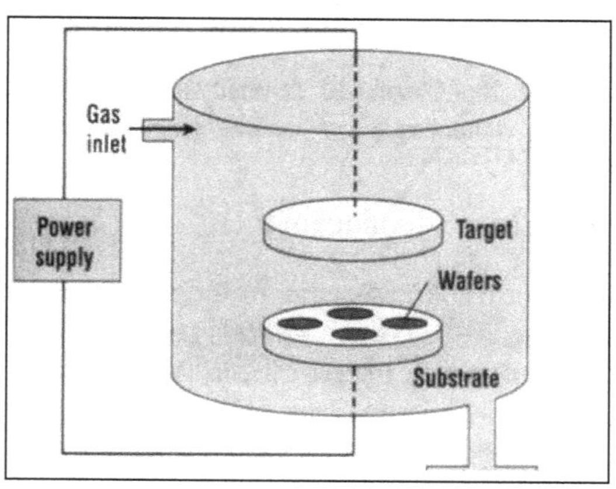

Sputtering.

- Sputtering: Argon ion bombardment of the material release particles of the material. Mainly for depositing rare earth magnets. Deposition rate and film surface area depend on sputtering tool and target size.

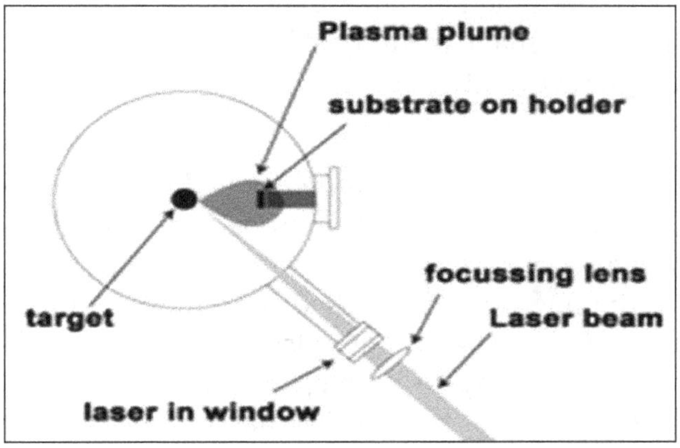

Pulsed Layer Deposition

- Pulsed Layer Deposition: a high-power pulsed laser beam is focused inside a vacuum chamber to strike a target of the material that is to be deposited.
- Electroplating.
- Screen printing.
- Wax/parylene bonding.

Issues with Magnetic Actuation

- High-power dissipation. This is a major problem for magnetic MEMS, but work is underway to circumvent this.
- Fabrication of the coil.
- Integration of the micromagnet into the MEMS device.
- Process-material compatibility.
- Integratability into the overall microfabrication process (maintain cost and throughput).
- So that preexisting processes in the fabrication of the MEMS device will not be tampered with, deposition temperatures and post-deposition treatment/conditions must be tolerable. Also, the micromagnet must be able to withstand any chemical treatment that will come after its deposition.
- Issues with magnetization (One may want to have more than one direction of magnetization; this creates a problem).

Each of these challenges can be mitigated or lessened by the right choice of material, choice of molding or fabrication method, and the type of device that is to be constructed. Applications of the magnetic actuator include: the synthetic jet actuator, micro-pumps and micro-relays.

Mini/Micro Shape Memory Actuators

Shape Memory Alloys (SMAs) are functional materials, characterized by some attractive features, such as pseudo-elasticity (PE) and shape memory effect (SME) . The latter consist in the capability of the material to recover high deformation values by heating and can be considered very suitable for actuation applications.

In the actuation field, the most common shape memory alloy is quasi-equiatomic NiTi system, commercially known as Nitinol®. Ti-rich NiTi compounds show characteristic transformation temperatures higher than the room one and they can recover high values of deformation. Moreover, these intermetallic compounds are widely used since they exhibit high thermal and mechanical cycling stability. Other NiTi-based alloys are also employed for this kind of applications. In particular, NiTiCu, with Cu substituting Ni in the 3-10at% range, or NiTiCo system are used respectively when the application requires narrow thermal hysteresis or high stiffness.

In figure, the comparison between differential scanning calorimetry (DSC) data derived from NiTi wire (200 μm in diameter) in fully annealed condition and after aging (500 °C for 10 minutes) are reported. As it can be seen, the fully annealed NiTi alloy goes into a one-step transformation both during cooling and heating. In this case, shape memory effect occurs by a martensitic transformation (MT) between a low-temperature monoclinic structure, B19', called martensite, and a high-temperature body-centered cubic parent phase, B2, called austenite. When the material is aged at specific temperatures, this transition may occur in two steps in association with a trigonal phase, called R-phase.

During heating, the phase transformation, named inverse MT, is defined by austenitic start (As) and finish (Af) temperatures. Similarly, during cooling direct MT is defined by martensitic start (Ms) and finish (Mf) temperatures, respectively.

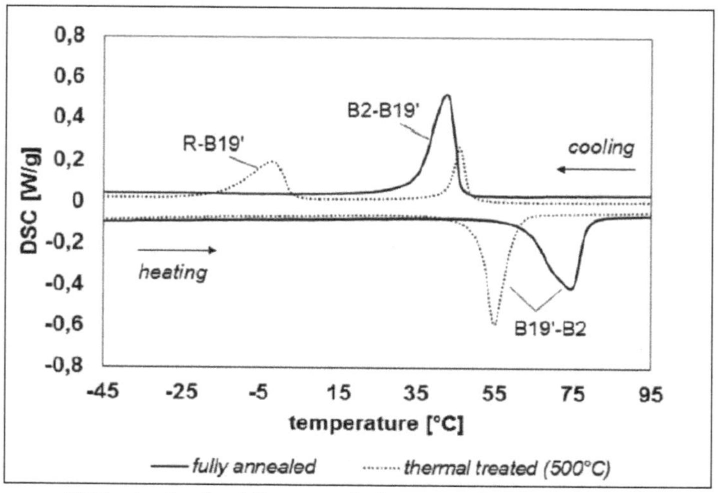

DSC scan of 200μm NiTi wire in the fully annealed state and after aging (500 °C for 10 minutes).

Then, the deformation induced by a state of stress on a SMA element can be recovered heating the material over the austenitic characteristic temperatures: the shape recovery starts at As and finishes at Af.

The transition temperatures are strongly influenced by the applied load, since it stabilizes the martensitic structure causing an increase of all transition temperatures. This behavior of SMA is well described by the Clausius-Clapeyron law.

Furthermore, SMA actuators generally need a training to stabilize their fatigue properties in terms of transformation temperatures and strain recovery. For example, a possible procedure consists in the thermal cycling under a constant load. Figure depicts the evolution of the temperature-stroke relationship during the first thermo-mechanical cycles. It is seen that the stabilization of the material behavior can be reached after few thermal loops.

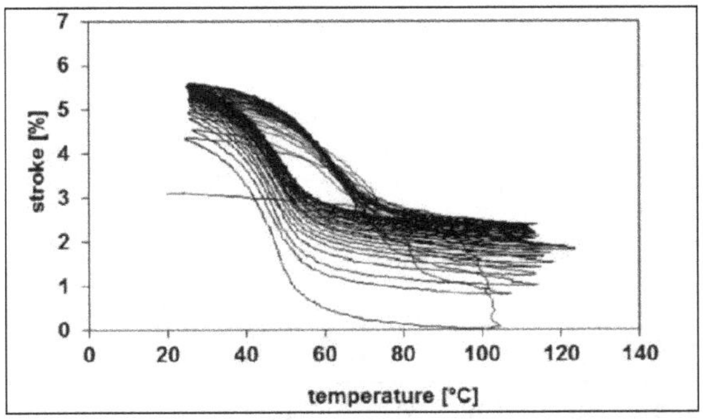

Example of thermo-mechanical hysteresis typical of a SMA material.

Since SMA can generate mechanical work, they are employed as actuators. The functional properties of SMA element used in actuators strictly depend on several factors, such as the chemical composition and the thermo-mechanical history.

Nowadays, the tendency is to produce ever smaller actuators because of the continued miniaturization of all the industrial products. In general, a mini-actuator may be composed by only a SMA element or by a mini-modular device, in which the SMA element is embedded and acts as the core and active part.

SMA Linear Actuators

The most important functional properties, which characterize shape memory wire for actuators, are the capability to recover high strain values, the characteristic transformation temperatures and the thermo-mechanical cycling stability . These characteristics of SMA can be tailored acting on chemical composition, thermal and mechanical processing, shape setting treatment and training In recent years, the good ductility of NiTi intermetallic has been exploited for the production of few tens micron wires.

Thanks to their small dimensions, such thin wires permit to reach very short cooling time but, obviously, they can be employed just in all those applications in which high loads are not required. The fabrication of these products has led to the miniaturization of SMA components for actuators that are nowadays employed for new kind of devices such as optical image stabilizer and autofocus for small camera.

During working life, the SMA wire is subjected to thermo-mechanical cycling and accumulates plastic deformation, elongating irreversibly and reducing its diameter. This behavior has to be considered in actuator design. Another important point to consider is the unsteadiness of the functional properties of shape memory wires during the working life. The plastic deformation induced during the thermo-mechanical cycling leads to a change of microstructure of SMA wire. Lattice defects, dislocations and nano-scaled precipitates are introduced into the matrix and they cause significant changes in transformation temperature and in the capability of the material to recover a deformation, as depicted in figure.

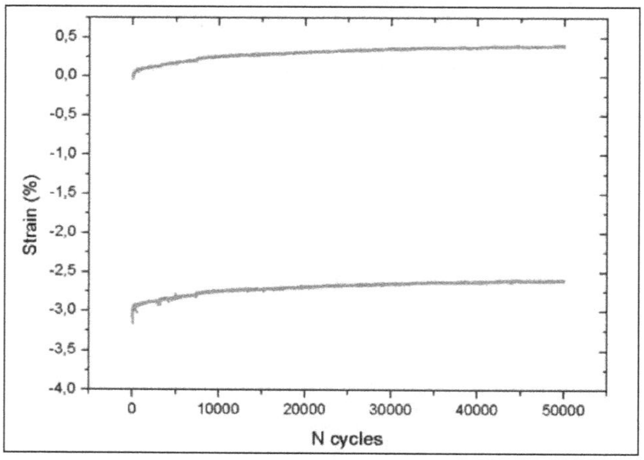

Thermal cycling of a 80μm wire under constant stress (200MPa).
The heating was performed by electrical pulse, and stroke was limited at 3%.

In most cases, SMA wires used as actuator, are heated by means of an electrical pulse (Joule effect) and cooled by natural air convection. So, it is very easy to reach short actuation time (heating time) but the reset time (cooling time) cannot be substantially controlled. Slight improvement in the reduction of reset time is achieved positioning the shape memory wire horizontally, as shown in figure. In this figure, it can be noticed that after heating (at 200s) the strain is maintained to two constant values for both the two configurations till time is near to 230s and 250s for horizontal and vertical position respectively. After these points, the two wires start the cooling process which results to be faster for the horizontally positioned wire than the vertical one. As an example, it can be seen that at 275s the horizontally positioned wire recovers half of its maximum strain while at the same time the vertical positioned wire has just started the cooling route.

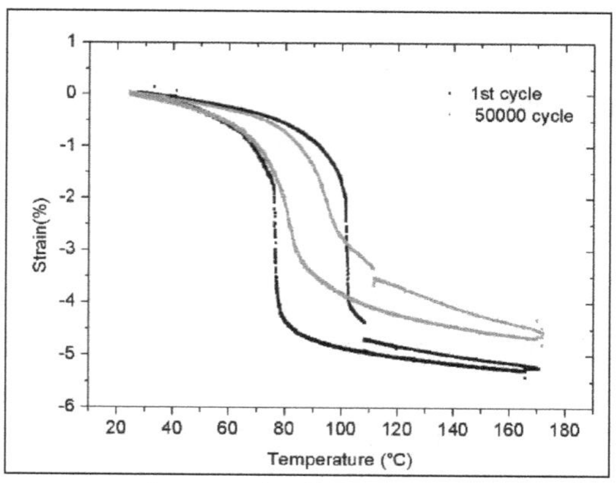

Thermal loop under constant stress (200MPa). The heating was performed by a thermal chamber.
The test was carried out on a 80μm wire before and after 5x104 thermal cycles under constant stress (200MPa).

As previously stated, the heating process of a SMA actuator can be easily obtained by Joule effect. In this case, it has to be considered that the shape of the electrical pulse used for the actuation strongly affects the functional properties of SMA. As an example, figure shows the time required by the wires to recover the 3% of deformation (actuation time) when it is heated by two different current pulses. At the 1st cycle ramp and step electrical pulse employed to heat two 80μm wires were designed to have the same electrical efficiency but the actuation time is 400ms by step current pulse while it is 623ms by ramp. After $5 \cdot 10^3$ cycles the wire heated by step pulse employs 618ms, so 218ms more than the time employed at the first cycle. As opposite, the ramp pulse leads to an increase of just 22ms. After $5 \cdot 10^4$ cycles the actuation time related to the wire heated by step is even higher than the one related to wire heated by ramp. This behavior leads to a drastic decrease of the step heating method efficiency (from 0.91% to 0.55%), vice versa the efficiency of the ramp heating method does not change so much (from 0.91% to 0.75%).

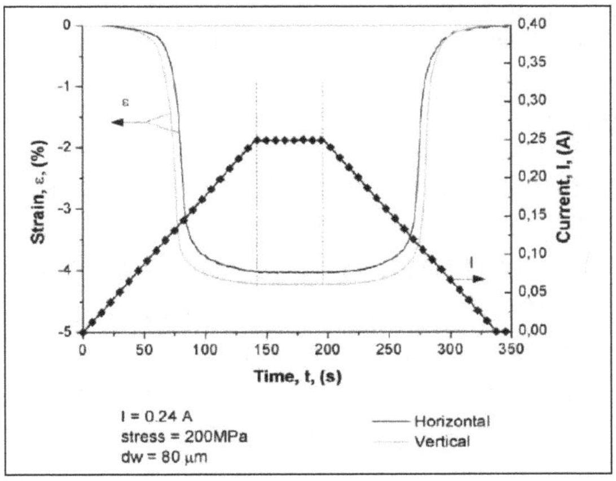

Comparison between the mechanical performance of horizontal and vertical configuration of 80μm SMA wire used as actuator.

Then, in order to achieve fatigue performance acceptable for the specific device, the right electrical pulse has to be chosen considering the number of cycles that the SMA actuator has been designed to work.

Recently, the effect of drawing procedure on functional fatigue of thin NiTi wires has also been investigated . Basically, 80μm NiTi wires were produced through two different drawing procedures reaching the same final cold working level before shape setting. These two processes differ for the number of drawing steps carried out to reach a certain cold working level before each heat treatment. After the last thermal treatment the specimen that underwent to a less number of drawing steps shows a narrow thermal hysteresis, even after thermo-mechanical cycling. It means that adopting a severe drawing procedure leads to improve the actuation performance.

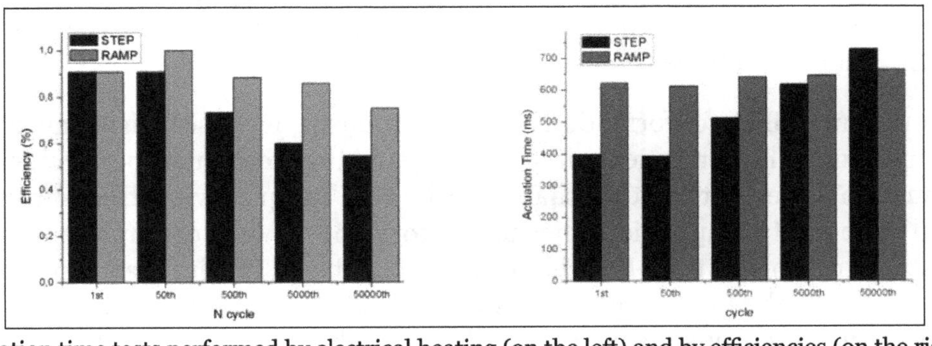

Actuation time tests performed by electrical heating (on the left) and by efficiencies (on the right).

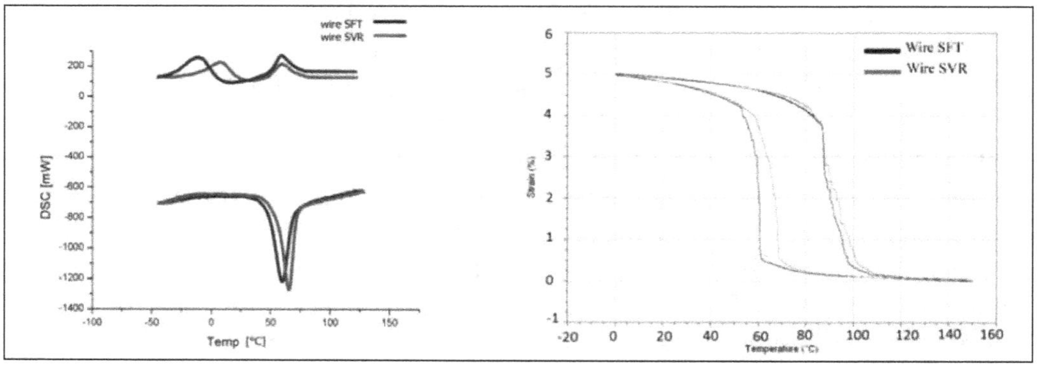

DSC scans, on the left, and thermal loop under constant tensile stress, on the right, of two 80μm NiTi wires obtained by a severe (SVR) and soft (SFT) drawing procedure.

Snake-like SMA Actuators

The main focus of miniaturization is to develop devices with high mechanical performance in very limited spaces. In the SMA field, the mini-actuator may be composed by only the SMA element or by a mini-modular device in which the SMA element is embedded and acts as the core part. In the latter case, some issues typical of SMA miniaturization should be taken into a serious consideration: first of all, the fixing points of the SMA element and the electrical contacts should be adequately designed to prevent abrupt rupture, and then the use of a wire geometry generally implies the employment of additional mechanical parts which may affect the long run performance of the SMA element.

The mechanical work resulting from the SMA shape recovery is used to produce linear or rotational motion in smart actuators. To attain high strokes in limited space, designers developed several miniature devices which include SMA elements with a straight or a helical spring shape . These two shapes present two opposite mechanical behaviors: the straight conformation allows high forces but smart wire arrangements should be employed in order to achieve also high strokes. As an example, Jansen et al. used two sets of seven pulleys to arrange a 1000mm long SMA wire with a diameter of 0.200mm, and attain 3% of stroke under a constant load of about 130MPa . The use of additional mechanical components may affect the overall performance of the SMA element, leading to friction which could influence the final functional parameters of the device. As opposite, the helical spring shape guarantees high strokes without the use of any further mechanical part but fairly small forces can be attained especially when embedded in miniature devices.

Hence, straight and helical spring conformations represent two opposite limits in the force versus stroke plane. In order to answer both to the need of stroke and force between these two limits, new SMA geometries should be studied.

In a recent work, a new SMA conformation has been proposed; it consists in a planar wavy formed NiTi wire, named snake-like arrangement. To understand the mechanical performance of this unusual conformation as respect to other shapes, three NiTi samples with different geometry, but with identical elemental composition, thermal history and power consumption were tested by means of a standard tensile test. In this investigation the NiTi wire was formed to have the straight conformation, the helical spring and the snake- like arrangements, and austenite and martensite mechanical responses were assessed separately.

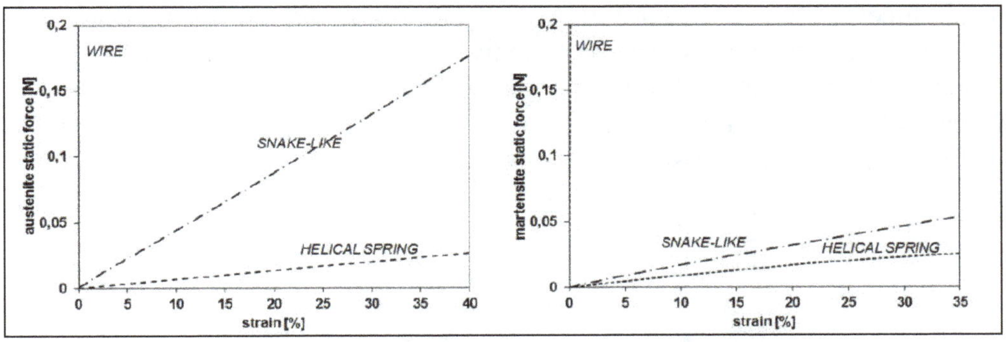

Comparison between austenite and martensite mechanical behavior of different NiTi wire geometries for actuators.

The snake-like conformation is characterized by four main parameters: the number of curvatures, N, the curvature radius, R, the distance between two consecutive curvatures, D, and the height, H. A snake-like specimen can be prepared by constraining a SMA wire in the snake-like arrangement by using a drilled aluminum bar and a series of nails; heating at high temperature is finally used to fix the shape. Figure reports a snake-like sample prepared from a commercial NiTi wire having the diameter of 0.2mm, formed at 500 °C during ten minutes and quenched in water at room temperature, with N=4, H=5.27mm±0.07mm, R=0.54mm±0.03mm and D=0.82mm±0.06mm.

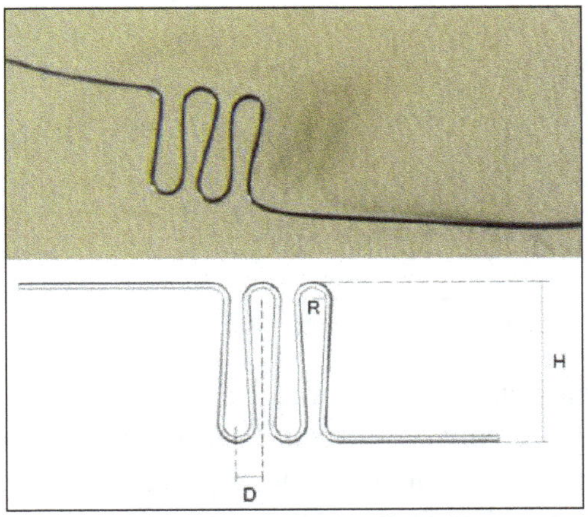

Snake-like NiTi wire with N=4, H=5.27mm±0.07mm, R=0.54mm±0.03mm and D=0.82mm±0.06mm.

As previously stated, during the ten minutes formation process the curved parts of the snake element are in contact with metal devices. This fact does not modify the calorimetric properties of the SMA element in all its length, as can be seen in figure where the overlapping of the DSC scans relative to the straight and the curved segments is clearly visible.

For small deformation values, this snake-like shape shows an almost complete recovery of the deformation both for austenite and for martensite.

The relative higher output stroke performed by the snake-like NiTi wire as respect to the straight wire conformation is substantially due to the bend/unbend movement of the curvatures. In order to avoid plastic deformation and to achieve simultaneously a significant output stroke, the snake sample, formed as previously explained, should be stressed with a load near 0.1N, as forces lower and higher than 0.1N do not guarantee a proper functioning of this snake element: in fact, lower forces do not induce martensite formation, as opposite higher forces cause plastic deformations. As an example, in figure the curves derived from five thermal cycles of the snake-like NiTi sample under two constant loads, 0.1N and 0.2N is depicted. By comparing the two trends, it can be seen that in both cases the snake sample lose part of its shape recovery during cycling; this is much more visible for the specimen stressed with 0.2N.

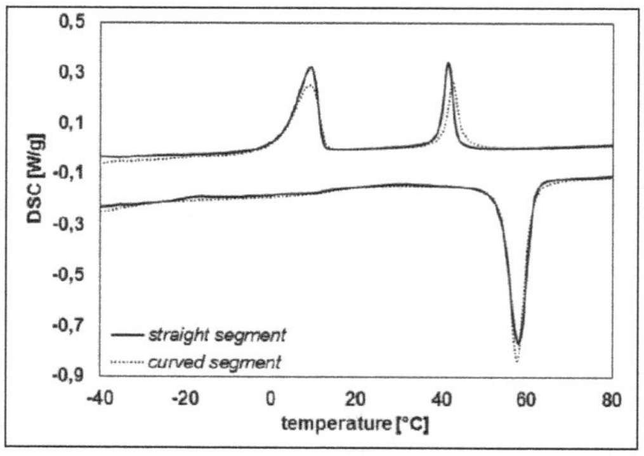

DSC scans of the straight and the curved part of the snake-like NiTi wire.

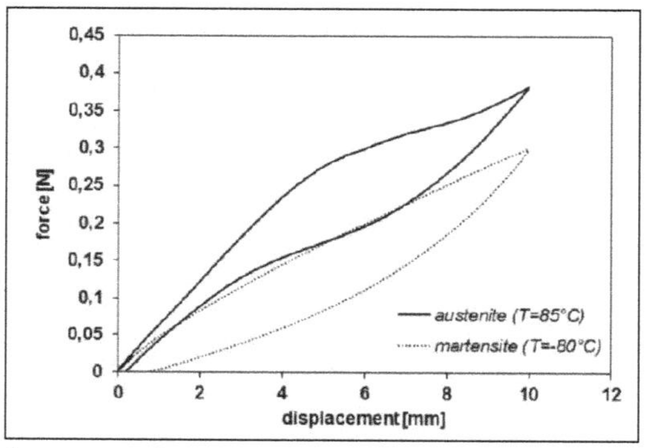

Force versus displacement of austenite and martensite of the snake-like NiTi wire.

Fatigue life was studied under the constant load of 0.1N; the snake-like sample was electrically activated with 0.6A with a maximum power consumption of 1W and a work frequency of 0.1Hz. Fracture takes place at the apex of that curvature most far from the applied load and from figure it can be observed that it occurs near the 7,5x10^4 cycle. The fractured section was analyzed through SEM observation, it can be seen a region with the striation of the crack propagation departing from a nucleation site placed on the inner surface of the snake and a region characterized by dimples, typical of ductile materials.

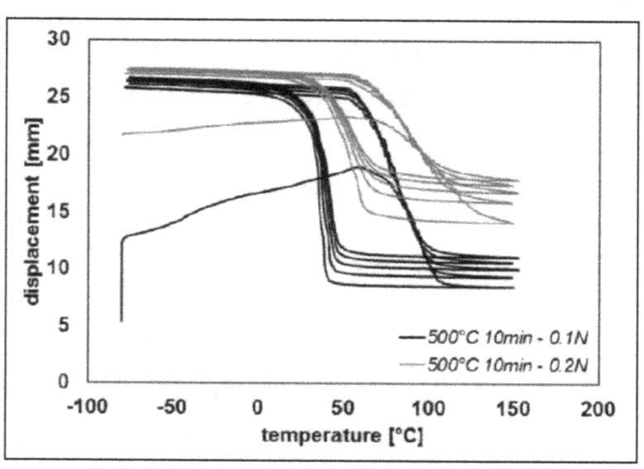

Hysteresis curves of the snake-like NiTi sample treated at 500 °C for 10 minutes under 0.1N and 0.2N constant load.

After the cycling test, the DSC scan shows a feeble shifting of the transformation temperatures towards higher values; a better definition of the rhombohedral phase during heating is also visible.

When embedded in mini-modular device, the snake-like wire does not lose its mechanical performance. As an example, in figure a miniature rotational actuator composed by two mutual antagonist NiTi snake-like wire is reported. The device exerts a stroke of about 120° under a constant torque of 0.1Nmm, as shown in figure; it is activated by 0.6A with a maximum power consumption of 1W and a work frequency of 0.05Hz. Due to the particular construction, plastic deformation of the SMA element is prevented as the SMA elements never exceed 6mm of deformation.

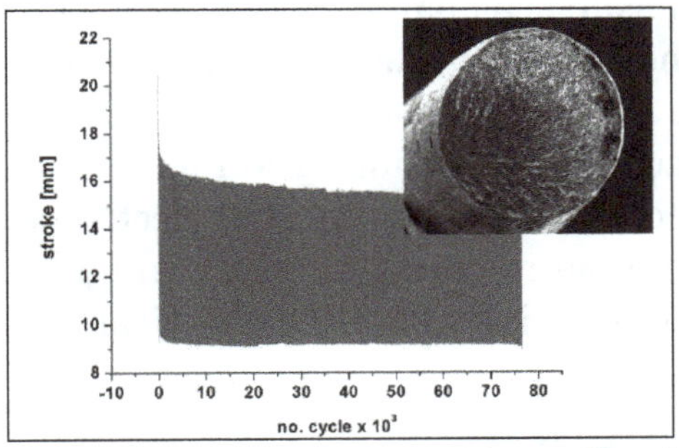

Snake-like NiTi sample fatigue test result under 0.1N load.

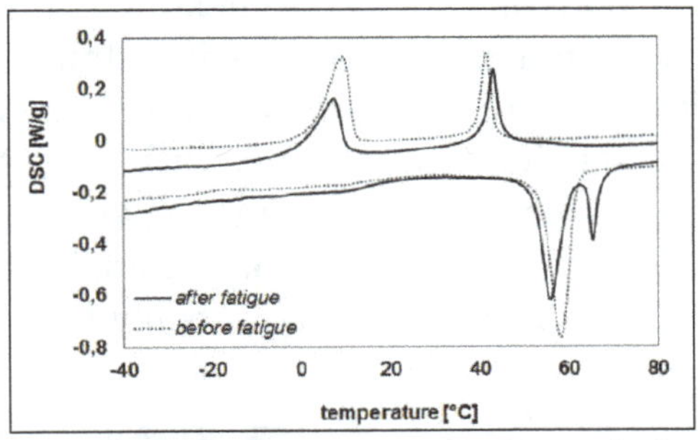

Snake-like NiTi sample DSC scan before and after the fatigue test.

Mini rotational actuator activated by two antagonists snake-like NiTi wires.

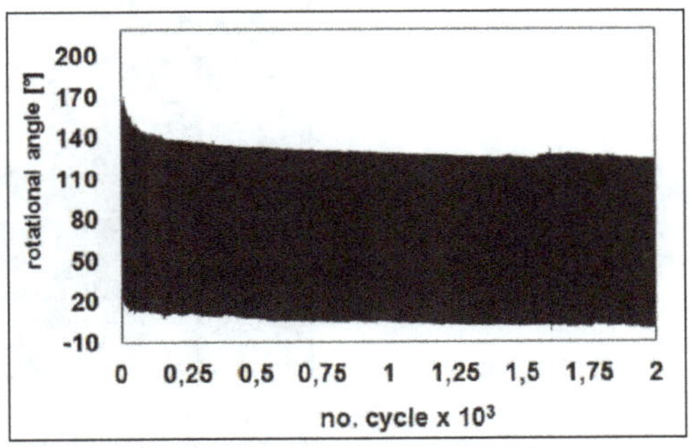

Fatigue test results of the mini rotational actuator under 0.1Nmm constant torque.

Micro Snake-like Shape Memory Alloy Elements

In the last years the trend of miniaturization of components and products has been evidently introduced in several industrial fields, such as biomedical, electronics, aerospace and mechanics. SMAs, as it already happens in biomedical applications can be machined using laser machining with appreciated and high qualitative results. The decision to investigate this method for the production of SMA actuators is considered an interesting topic because it shows some relevant advantages in comparison with other technologies, such as high productivity, high quality and repeatability, when applied in the industrial world.

Hence, the snake geometry can be easily scale down to the micro scale by the employment of laser technology with appreciated and high qualitative results. Because laser material processing is a thermal process, in which an heat source is used to remove a certain volume of material from the work piece, some thermal damages are evidently obtained, such as melted material, heat affected zone and oxides. To remove these defects, chemical etching and electro-chemical polishing have to be performed on the snake SMA samples. Figure depicts SEM images of the snake NiTi sample after laser, chemical etching and electro-chemical processes; top and bottom surfaces as well as snake sample section are shown. It can be seen that laser damages are mainly visible on the bottom surface and that the chemical etching remove the great part of defects. Moreover, a significant loss of mass and consequent reduction of the dimensions of the micro-snake can be evidently observed from SEM pictures due to the material removal of the post-processing. The snake NiTi sample geometry and calorimetric properties change after each process. In particular, the geometrical dimension are referred to a micro-snake obtained starting from a 120μm NiTi sheet. As concern the DSC data, it can be observed that the laser machining produces a refinement of the rhombohedral phase peak both on cooling and heating while the martensite peak barely change as respect to the unworked 120mm thick sheet, being almost spread and invisible. The laser process also yields a conspicuous diminution of the rhombohedral phase transformation temperatures and of the martensite starting temperature while the other characteristic transformation temperatures do not have an analogous variation being almost unchanged. Chemical etching and electro-polishing steps do not visibly alter any phase transformation temperature both on cooling and on heating, even if some little change due probably to the hydrogen introduction in the chemical process, is visible.

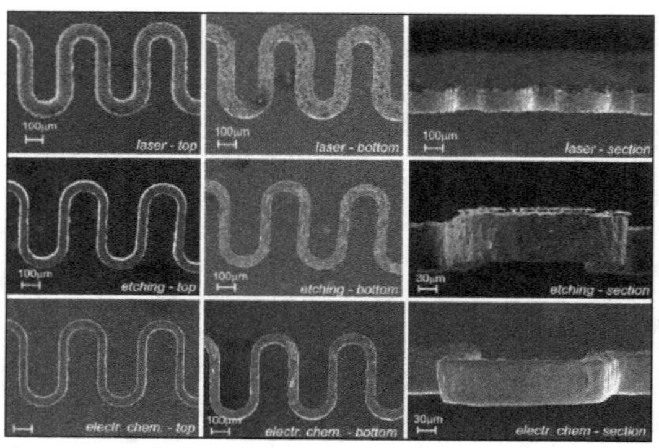

SEM images of the snake NiTi sample surfaces and section after laser, chemical etching and electro-polishing.

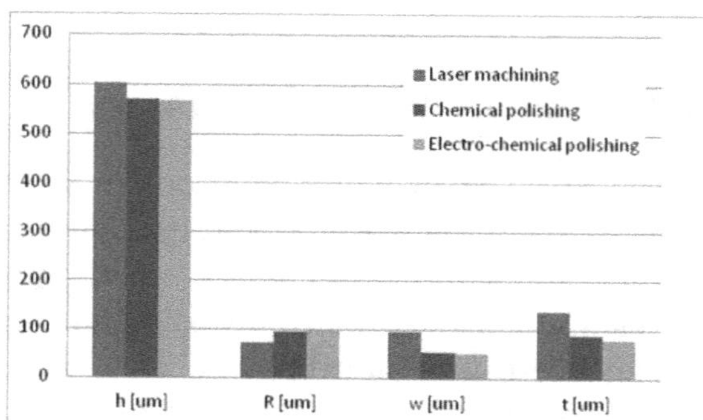

Processes Variation of the geometrical parameters of the micro-snake during the fabrication process (h: high, R: radius of curvatures, w: width, t: thickness).

The fabrication process previously described can be used to produce snake SMA sample of different length. As an example, figure depicts the SEM image of a micro-snake NiTi sample with 19 curvatures (i.e. 3.8mm of length) obtained starting from a 120μm NiTi sheet treated at 400°C for 15 minutes and quenched in water.

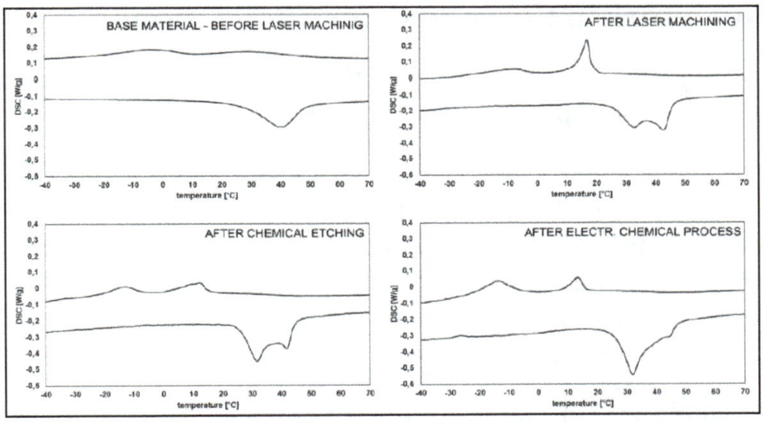

DSC scans before laser machining and in correspondence of each fabrication process.

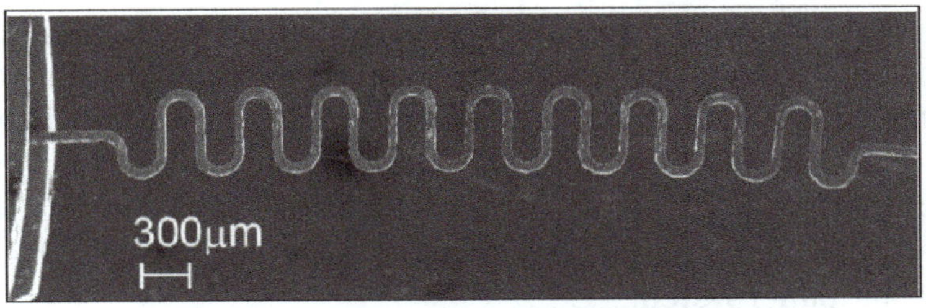

SEM image of a micro-snake NiTi sample obtained starting from a 120μm NiTi sheet.

Strain-recovery and thermo-mechanical cycling tests demonstrate the high performance of this micro-snake sample. As an example, under the constant load of 16mN, it reaches about 40% of recovery with a cycling stabilization within few hundreds of cycles, when heated by electrical current (130mA) with a maximum power consumption of 0.3W and a frequency of 0.15Hz.

Industrial Applications of Electric Linear Actuator

Electric linear actuators are among the most useful inventions till date. These equipment have made so many jobs easy and efficient. These actuators are known for their wide range of applications in a variety of industries, which include military, agriculture, food, material handling, and many more. Their compatibility with so many industries make them versatile, and popular.

Uses of Linear Actuators in the Industrial Applications

There are a number of industrial applications, which employ electric linear actuators. Here are some most common and major application areas of these actuators:

- Window Automation: This is one of the most talked about areas of application of electric linear actuators. Actuators have a vital role to play in window automation. They allow for easier access and an improved ventilation. How? Actuators help in automating the opening and closing of windows for natural ventilation. This natural ventilation is useful for keeping the temperature of the building low at night, and even in extracting harmful smoke and heat during any fire emergency. This is only possible with the help of electric actuators.
- Agriculture & Outdoor: Agricultural industry implements several heavy duty equipment, such as harvesters, spreaders, tractors, and many more. These equipment are required to carry out linear motion to operate. Linear actuators have the capability to withstand extreme work conditions in the agricultural open field. Combine harvesters use levers to divert the grain flow. These levers are operated with the help of linear actuators. Additionally, the nozzle angle of the spreaders can be adjusted by these actuators. Other than agriculture industry, electric actuators are also used for solar tracking applications. The angle of the solar panels can be adjusted to follow the sun's orbit, which helps increase the efficiency of solar panel system.
- Special Factory Machinery**:** Electric linear actuators are being used increasingly in manufacturing industries for automating equipment, such as material handlers, slurry tanks, and several automotive equipment. Applications, such as factory machinery used to cut carpet rolls also employ linear actuators for raising or lowering the blades.

These were some major industrial application areas, which use the electric linear actuators extensively. Employing these actuators in these applications increases the overall efficiency, and betters the performance. However, the manufacturer from whom you buy a linear actuator matters a lot. It is always a good idea to consult experts and get these extremely useful equipment from trusted and experienced manufacturers and suppliers. This will not only help you get good quality actuators, but also improve the overall performance.

References

- Ponomareva, Papers, courses, Jasso, konferenzen: in.tum.de, Retrieved 16 February, 2019
- T. Rueckes; et al. (2000). "Carbon Nanotube-Based Nonvolatile Random Access Memory for Molecular Computing". Science. 289 (5476): 94–7. Bibcode:2000Sci...289...94R. Doi:10.1126/science.289.5476.94. PMID 10884232
- Actuators, piezo-theory: americanpiezo.com, Retrieved 16 February, 2019
- Chin, Tsung-Shune (2000). "Permanent magnet films for applications in Micro-electro-mechanical systems". Journal of Magnetism and Magnetic Materials. 209 (1): 75–79
- Introduction-to-piezoelectric-actuators: smashingrobotics.com, Retrieved 17 March, 2019
- Chin, Tsung-Shune (2000). "Permanent magnet films for applications in Micro-electro-mechanical systems". Journal of Magnetism and Magnetic Materials. 209 (1): 75–79. Bibcode:2000JMMM..209...75C. Doi:10.1016/S0304-8853(99)00649-6
- Electrical-operation, properties-piezo-actuators, piezo-technology: piceramic.com, Retrieved 18 April, 2019

PERMISSIONS

All chapters in this book are published with permission under the Creative Commons Attribution Share Alike License or equivalent. Every chapter published in this book has been scrutinized by our experts. Their significance has been extensively debated. The topics covered herein carry significant information for a comprehensive understanding. They may even be implemented as practical applications or may be referred to as a beginning point for further studies.

We would like to thank the editorial team for lending their expertise to make the book truly unique. They have played a crucial role in the development of this book. Without their invaluable contributions this book wouldn't have been possible. They have made vital efforts to compile up to date information on the varied aspects of this subject to make this book a valuable addition to the collection of many professionals and students.

This book was conceptualized with the vision of imparting up-to-date and integrated information in this field. To ensure the same, a matchless editorial board was set up. Every individual on the board went through rigorous rounds of assessment to prove their worth. After which they invested a large part of their time researching and compiling the most relevant data for our readers.

The editorial board has been involved in producing this book since its inception. They have spent rigorous hours researching and exploring the diverse topics which have resulted in the successful publishing of this book. They have passed on their knowledge of decades through this book. To expedite this challenging task, the publisher supported the team at every step. A small team of assistant editors was also appointed to further simplify the editing procedure and attain best results for the readers.

Apart from the editorial board, the designing team has also invested a significant amount of their time in understanding the subject and creating the most relevant covers. They scrutinized every image to scout for the most suitable representation of the subject and create an appropriate cover for the book.

The publishing team has been an ardent support to the editorial, designing and production team. Their endless efforts to recruit the best for this project, has resulted in the accomplishment of this book. They are a veteran in the field of academics and their pool of knowledge is as vast as their experience in printing. Their expertise and guidance has proved useful at every step. Their uncompromising quality standards have made this book an exceptional effort. Their encouragement from time to time has been an inspiration for everyone.

The publisher and the editorial board hope that this book will prove to be a valuable piece of knowledge for students, practitioners and scholars across the globe.

INDEX

www.ingramcontent.com/pod-product-compliance
Lightning Source LLC
LaVergne TN
LVHW080707210926
848485LV00003B/429
* 9 7 8 1 6 4 7 2 8 3 2 8 5 *